Feature Management with LaunchDarkly

Discover safe ways to make live changes in
your systems and master testing in production

Michael Gillett

BIRMINGHAM—MUMBAI

Feature Management with LaunchDarkly

Group Product Manager: Aaron Lazar
Publishing Product Manager: Harshal Gundetty
Senior Editor: Ruvika Rao
Content Development Editor: Vaishali Ramkumar
Technical Editor: Pradeep Sahu
Copy Editor: Safis Editing
Project Coordinator: Deeksha Thakkar
Proofreader: Safis Editing
Indexer: Sejal Dsilva
Production Designer: Vijay Kamble

First published: September 2021
Production reference: 1230921

Published by Packt Publishing Ltd.
Livery Place
35 Livery Street
Birmingham
B3 2PB, UK.

ISBN 978-1-80056-297-4

www.packt.com

To my family, for their constant advice and support, without which I would not be where I am today. I hope this book sheds some light on what I do in my job.

– Michael Gillett

Foreword

Writing a foreword to a book about using the software of the company you cofounded veers dangerously close to "going to a concert and wearing the band's t-shirt" levels of cheerleading. The opportunity here is worth it though – as the CTO and cofounder of LaunchDarkly, I've seen firsthand the transformative impact feature management can have on an organization shifting to a digital-first world. Engineering leaders have told me that feature management has revolutionized the way they deliver software. I've seen data quantitatively demonstrating the positive change feature management has on the speed and quality of software delivery in hundreds of organizations.

I've also seen the exact opposite reaction to the idea of feature management. Transformative impact? Revolutionary change? We're talking about feature flags. They're just `if` statements in code, right? Why would such a simple concept need a third-party tool or a full-fledged book? One of my favorite Hacker News comments about LaunchDarkly sums up the sentiment well – "Isn't this just… a hash table?" Or, even better: "Is there really a venture-backed company out there doing booleans as a service?"

What I've come to learn from founding LaunchDarkly is that both sides are right – and that's what makes feature management worth learning about. Feature flags are incredibly simple, and yet, they can transform the way you develop software.

There's no better way to learn this than from someone that's lived it firsthand.

Michael was one of the earliest customers of LaunchDarkly, having started using the platform in 2016. As such, he brings a unique perspective to his book – that of a long-time user who's been along for the entire journey, from the decision to bring in a feature management solution, to initial adoption, to widespread rollout and steady-state use. This perspective is what makes his book so powerful – Michael's writing combines his many years of experience with the LaunchDarkly platform with his deep understanding and insight into the value of feature management.

Michael's book is a helpful guide for anyone considering or currently adopting feature management. If you're at the beginning of your journey, you'll find answers to your questions about the business value of feature management. You'll get a tour of the most common use cases for feature management – from mitigating risk with kill switches to building a practice of experimentation. You'll discover what capabilities you need in a feature management platform upfront – not just today's needs, but the needs that you'll eventually grow into as your practice evolves. Finally, you'll learn how to best set up and structure the LaunchDarkly platform for your organization from the outset – following best practices learned over years of experience.

Michael's book is one of the most complete resources on LaunchDarkly that I've seen. I'm confident that after reading his work, you'll come to the same conclusion that I have – that feature management is both incredibly simple and incredibly powerful, and will soon be a standard practice for any modern software development team.

John Kodumal

CTO and cofounder of LaunchDarkly

Contributors

About the author

Michael Gillett is a head of development and a full-stack software engineer residing in London, UK. He has worked with feature management and LaunchDarkly for several years and has defined processes and techniques to enable teams to get the most from this approach to software delivery. He often talks on this subject at conferences and events.

Michael graduated from the University of Hertfordshire with an MSc in computer science and a BEng in electrical and electronic engineering in 2012. Since 2012 Michael has been a Microsoft MVP, currently in the Windows Insider category.

I want to thank all those who have supported and encouraged me throughout the process of authoring this book, especially my parents and brother. I want to thank all those at Packt for this opportunity and for your help and feedback during the development of this book, including the technical reviewers, Dawn and Daniel.

About the reviewers

Dawn Parzych is a developer marketing manager at LaunchDarkly where she helps developers ship code fast and safely. She specializes in the socio-technical aspects of technology and how humans are often the hardest part. Her career in tech has included roles with companies such as Catchpoint and F5. Her mission is to make tech more accessible and inclusive by using straightforward language and fewer buzzwords. Dawn recently relocated from the Pacific Northwest to the Midwest and she's slowly adjusting to life without mountains and oceans.

Daniel Yefet has more than 10 years of experience in web development and is currently working as a lead developer in a London-based technology company. He focuses on web performance and experimentation using technology within the JavaScript ecosystem, such as Node.js and React. In his spare time, he loves spending time with his wife and kids, playing music, and coding!

Table of Contents

3

Basics of LaunchDarkly and Feature Management

Section 2: Getting the Most out of Feature Management

4

Percentage and Ring Rollouts

5

Experimentation

6

Switches

7

Trunk-Based Development

8

Migrations and Testing Your Infrastructure

Section 3: Mastering LaunchDarkly

9

Feature Flag Management in Depth

10

Users and Segments

Preface

Over the past few years, DevOps has become the de facto approach for designing, building, and delivering software, but feature management is now extending this methodology. From feature toggles to experimentation, from rollouts to migrations, if you want to make feature management happen, LaunchDarkly is the tool for you.

This book explains how feature management is key to building modern software systems. Starting with the basics of LaunchDarkly, you will learn how the simple feature flag can be used to turn features on and off. Then you will discover percentage- and ring-based rollouts, how to gain insights from your customers via experiments, and how switches can help maintain a good production environment. You will see how feature management can change the way teams work and how large projects, including migrations, are successfully achieved.

Finally, you will discover various uses of every part of the tool to gain mastery of LaunchDarkly. This includes tips and tricks for experimentation, identifying groups and segments of users, and investigating and debugging issues with specific users and feature flag evaluations. By the end of the book, you will have gained a comprehensive understanding of LaunchDarkly, along with knowledge of the adoption of trunk-based development workflows, multi-variant testing, and managing infrastructure changes and migrations.

Who this book is for

This book is for developers, quality assurance engineers and DevOps engineers. This includes individuals who want to decouple the deployment of code from the release of a feature, run experiments in production, or understand how to change processes to build and deploy software. Software engineers will also benefit from learning how feature management can be used to improve products and processes. A basic understanding of software is all that you need to get started with this book as it covers the basics before moving on to more advanced topics.

What this book covers

Chapter 1, Introduction to Feature Management with LaunchDarkly, explains the modern software development landscape, where **feature management** fits within it, and how feature management can be integrated
into CI/CD processes. This context will lay the groundwork for the following chapters to explain how feature management can be used to add value to products, teams, and businesses.

Chapter 2, Overview of Feature Management, details what feature management is, how it empowers development and product teams, and how it can be used to reduce risk. Examples are provided for the two main types of **feature flag** with a comparison between them. The chapter concludes with information about testing in production.

Chapter 3, Basics of LaunchDarkly and Feature Management, introduces **LaunchDarkly**, explaining how to implement the tool, detailing the functionality available to create and manage feature flags, and exploring the concept of LaunchDarkly's user system. Additionally, the chapter details the role management functionality of LaunchDarkly to allow you to get started in the right way.

Chapter 4, Percentage and Ring Rollouts, focuses on the **percentage** and **ring rollout** use cases for feature management. Examples of this include new implementations that need monitoring to validate that they are working as expected in production, through to rolling out a new feature to only a small group of customers and validating that their experience is good.

Chapter 5, Experimentation, explains the opportunities for **experimentation** and how to **test in production**. It looks at the types of experiments that can be performed and shows how LaunchDarkly enables teams to run effective experiments.

Chapter 6, Switches, explains the use of **switches**, how they are useful in certain situations, and how they can be managed by LaunchDarkly. The chapter also explores how LaunchDarkly can be used to manage permanent feature flags.

Chapter 7, Trunk-Based Development, explores one of the more extreme opportunities available to teams using LaunchDarkly, **trunk-based development**. The ability to safely **test in production** presents the opportunity to remove the need for feature branches and bypass any test environments entirely and this chapter looks at how to achieve this.

Chapter 8, Migrations and Testing Your Infrastructure, explains how infrastructure **migrations** can be performed using LaunchDarkly, especially to reduce the risk of them negatively impacting your systems. The chapter goes on to detail how your infrastructure can be tested in novel ways through LaunchDarkly.

Chapter 9, Feature Flag Management in Depth, details, in depth, the **feature flag** functionality within LaunchDarkly to empower you to get the most out of the tool. This covers topics from viewing and managing all your flags to being able to configure, target, and experiment with a single flag.

Chapter 10, Users and Segments, provides an overview of the **users** and **segment** functionality in LaunchDarkly. It details how to view, set up, and edit users and segments. Understanding how to target features to the correct customers is crucial to the success of feature management and this chapter covers how best to accomplish this.

Chapter 11, Experiments, shows how to set up and make full use of **experiments** in LaunchDarkly by detailing the types of metrics and events available in the tool. Being able to manage and gain insights from experimentation is essential for the practice of testing in production and this chapter shows all the ways in which LaunchDarkly aids in working with methodology.

Chapter 12, Debugger and Audit Log, details the **Debugger** section and how it can be used to view what is happening with flag, user, and experimentation events. Information is provided to show how powerful insight details can be. The chapter concludes with exploring the **Audit Log** section, which is useful to understand any changes within a LaunchDarkly project and/or environment.

Chapter 13, Configuration, Settings, and Miscellaneous., explores the **Account** section of LaunchDarkly, which provides functionality to manage the members of the team, including **roles** and **permissions**. It details the security features that can be enabled to control access to the tool, which is especially important given how powerful it is. The chapter also looks at the **billing** and **usage** information provided by LaunchDarkly, before concluding with a look at the **profile** functionality and **history** logs. While this functionality does not directly impact the feature manager, it is essential for good governance and offers the flexibility for the tool to be configured to suit any business' needs.

To get the most out of this book

In some of the chapters, there are step-by-step code samples for how to implement LaunchDarkly and some of its functionality. There are sample applications in a GitHub repository that complement these code samples, providing a blank application or a complete one to enable you to follow along and test out the implementations yourself. The sample applications make use of the following technology:

Software used in the book	Operating system used for the samples
Git	Windows
Visual Studio 2019 or Visual Studio Code	
.NET 5	
LaunchDarkly SDK	

The code samples were written and tested on Windows. However, as .NET 5 is a cross-platform framework, the samples should work on Linux and macOS.

If you are using the digital version of this book, we advise you to type the code yourself or access the code from the book's GitHub repository (a link is available in the next section). Doing so will help you avoid any potential errors related to the copying and pasting of code.

You may benefit from following the author on Twitter (`https://twitter.com/michaelgillett`) or connecting with them on LinkedIn (`https://www.linkedin.com/in/michaelwgillett/`).

Download the example code files

You can download the example code files for this book from GitHub at `https://github.com/PacktPublishing/Feature-Management-with-LaunchDarkly`. If there's an update to the code, it will be updated in the GitHub repository.

We also have other code bundles from our rich catalog of books and videos available at `https://github.com/PacktPublishing/`. Check them out!

Download the color images

We also provide a PDF file that has color images of the screenshots and diagrams used in this book. You can download it here: `https://static.packt-cdn.com/downloads/9781800562974_ColorImages.pdf`.

Conventions used

There are a number of text conventions used throughout this book.

`Code in text`: Indicates code words in text, database table names, folder names, filenames, file extensions, pathnames, dummy URLs, user input, and Twitter handles. Here is an example: "Mount the downloaded `WebStorm-10*.dmg` disk image file as another disk in your system."

A block of code is set as follows:

```
public void OnGet()
       {
               User user = LaunchDarkly.Client.User.Builder(Guid.
                       NewGuid().ToString())
               .Anonymous(true)
               .Build();
       }
```

Bold: Indicates a new term, an important word, or words that you see onscreen. For instance, words in menus or dialog boxes appear in **bold**. Here is an example: "In this chapter, we will explore both the **Debugger** and **Audit log** sections and understand what information they impart."

> **Tips or important notes**
> Appear like this.

Get in touch

Feedback from our readers is always welcome.

General feedback: If you have questions about any aspect of this book, email us at `customercare@packtpub.com` and mention the book title in the subject of your message.

Errata: Although we have taken every care to ensure the accuracy of our content, mistakes do happen. If you have found a mistake in this book, we would be grateful if you would report this to us. Please visit `www.packtpub.com/support/errata` and fill in the form.

Piracy: If you come across any illegal copies of our works in any form on the internet, we would be grateful if you would provide us with the location address or website name. Please contact us at `copyright@packt.com` with a link to the material.

If you are interested in becoming an author: If there is a topic that you have expertise in and you are interested in either writing or contributing to a book, please visit `authors.packtpub.com`.

Share Your Thoughts

Once you've read *Feature Management with LaunchDarkly*, we'd love to hear your thoughts! Scan the QR code below to go straight to the Amazon review page for this book and share your feedback.

https://packt.link/r/1-800-56297-7

Your review is important to us and the tech community and will help us make sure we're delivering excellent quality content.

Section 1: The Basics

This first section provides an overview of the modern software development landscape and introduces where feature management and LaunchDarkly fit within it. The section provides examples of how feature management can enhance the DevOps methodology to make deploying code and releasing new features safer and more effective. The final topic that this section covers is how LaunchDarkly can be implemented in an application and how feature flags can be used.

This section comprises the following chapters:

- *Chapter 1, Introduction to Feature Management with LaunchDarkly*
- *Chapter 2, Overview of Feature Management*
- *Chapter 3, Basics of LaunchDarkly and Feature Management*

1

Introduction to Feature Management with LaunchDarkly

Welcome to this journey of **feature management** discovery. Throughout 13 chapters, we will explore what **feature management** is and learn how to use LaunchDarkly to get the most out of it. **Feature management** is the name given to the process of changing the elements of a system without making changes to an application's code base. This is not a new concept. However, there are modern ways to achieve feature management, and LaunchDarkly is a tool that enables our teams to make changes at runtime for specific applications, users, or all customers.

Feature management and LaunchDarkly empower us to change systems quickly, easily, and without risk to help deliver new features, deal with service incidents, and migrate systems.

In this chapter, I will introduce you to, and briefly explain, the modern software development landscape with regard to best practices for DevOps and robust **Continuous Integration/Continuous Delivery** (**CI/CD**) pipelines and processes. Next, we will look at where feature management fits within this practice and how it can be embedded effectively and efficiently within CI/CD processes.

This context will lay the groundwork for this book to explain what feature management is, how best to use LaunchDarkly, and what can be achieved by adopting an approach to building software that relies on feature management.

This chapter covers the following topics:

- Understanding modern software development
- Where does feature management come from?
- Feature management within CI/CD

By the end of this chapter, you will have a better understanding of how feature management fits within a modern approach to software delivery and how it can improve good DevOps and CI/CD practices.

Understanding modern software development

Recent years have seen the move to a more modern approach to software development, with DevOps being increasingly used across many industries. DevOps is best described by Microsoft as *the union of people, processes, and technology to continually provide value to customers.*

DevOps has ushered in the era of working where teams are more empowered than ever before, with ownership across the whole development life cycle of both code and products. This includes managing the production environment, previously the domain of the Operations team. The annual State of DevOps Report (available in the *Further reading* section; registration is required) shows that teams adopting DevOps are delivering higher quality code faster and more reliably than ever before.

While there are other approaches to working with software, DevOps is perhaps the most popular and has proven success. I will assume that this approach is how you are building and deploying software too. Before explaining feature management or even getting started with LaunchDarkly, I want to share a brief overview of DevOps to highlight where feature management can really add value.

Introducing the DevOps life cycle

DevOps is not just a practice of writing code but a comprehensive approach to defining, designing, and delivering software. It can be defined into four key areas within the application life cycle: **planning**, **developing**, **delivering**, and **operating**. I will describe these areas within the context of where feature management can be factored in later:

- **Planning**: Within this stage, the team works to understand the problems to solve, reviews the available data, and creates ideas and plans for features to improve their system and to add more value for their customers.

- **Developing**: In this stage, the features are implemented and checked against quality measures to ensure that the planned feature will work to deliver the expected functionality.

- **Delivering**: When in the delivering phase, the implemented feature goes through the build and release pipelines, ultimately to be *deployed to the production* environment to realize the expected value. Ideally, the release process uses **CI/CD pipelines** and is automated to an extent that is possible. These pipelines can often deploy the feature to several environments before it's deployed to production to validate that the system remains functional with no regressions.

- **Operating**: Once the feature is in production, it needs to be monitored and the data needs to be analyzed to ensure that the feature works as expected and delivers the planned value. This requires a mature implementation of monitoring, with the team also relying on this data to alert them to issues before they impact the customers. At this point, the data can be used as feedback in the *planning* phase to allow the team to identify new features and work to deliver additional value.

The aforementioned phases of the DevOps life cycle are highly effective at improving the processes of software delivery but aren't the whole practice of DevOps. Next, we will look at the culture that needs to go alongside these stages to make DevOps truly effective.

Introducing the DevOps culture

Successfully adopting a good DevOps practice is more than just following processes and procedures. It is also about adopting a culture that allows constant improvements to be made to the methodology. Microsoft defines the DevOps culture as having four key areas (available in the *Further reading* section). Again, I will explain only the areas that come within the scope of feature management:

- **Collaboration, visibility, and alignment**: For teams to be successful, they must be able to collaborate and communicate well to better understand and define problems and plan effective features and solutions. For teams to be able to trust one another, there needs to be an alignment of timelines and good transparency and visibility of the progress of the work being undertaken.

- **Shifts in scope and accountability**: I touched on this in the DevOps life cycle overview, where I stated that accountability and ownership must lie with the development teams. Often, this requires a shift from the traditional approach where developers, QA, and the Operations teams work separately to one where development teams are responsible for writing, testing, and deploying a system, along with running it in production. This allows teams to take a holistic view of the problem and solution and achieve the most value for their customers.

- **Shorter release cycles**: Being able to release new features faster is crucial in being able to iterate and gain more value quickly. It also reduces risks and improves stability throughout the whole DevOps life cycle. This encompasses all aspects of DevOps, from **planning** through to **operating**.

- **Continuous learning**: A *growth mindset* is the key to an effective DevOps workflow as it allows teams to question everything and learn what works best for their business and customers. This requires an appetite to fail fast and learn from those failures, which helps to improve the systems and processes so that you gain the most value from them. This is possibly the most important aspect of DevOps as without this, the entire DevOps life cycle breaks down. Without this, constant improvement and growth can not be achieved.

Next, we look into the four key DevOps metrics.

Introducing the four key DevOps metrics

It is important to understand that a common way to measure how mature a team's DevOps practice is is with the **four key metrics of DevOps** (these are defined in *State of DevOps 2019*, which is available in the *Further reading* section):

- **Deployment Frequency**: This measures how often deployments can be conducted. The ideal value for this metric is to be able to deploy on demand and multiple times a day.

- **Lead Time to Change**: This measure looks at how long it takes for a change to go from committed code to the production system. The target time is *less than 1 day*.

- **Time to Restore Service**: It is important to recover from a service incident quickly. A mature **DevOps** team should be able to recover from a production issue in *less than an hour*. This metric allows a team to see the average time it takes to restore normal services.

- **Change Failure Rate**: This measures the number of deployments that have an undesired impact on the production system. The accepted percentage of releases that result in a degraded service for a mature **DevOps** team is between *0 – 15%* of releases.

The best performing teams achieve their reliable results by following both the **DevOps life cycle** and by having a good **DevOps culture**.

By using DevOps, software teams and companies have transformed how effective they are, but this practice alone does not solve problems, nor does it limit how new ideas can make this even more effective. One way in which DevOps can be enhanced further is with the ability to manage features within systems and applications, which we will look at in the next section. We will also examine how feature management can positively impact the four key DevOps metrics.

Where does feature management come from?

I will explain what feature management is in detail in the next chapter, but before that, I will talk broadly about how this approach can be applied to a modern DevOps way of working and how it fits within the DevOps life cycle. The idea of **feature management** in modern systems is self-explanatory – it is an approach to being able to manipulate components within a production system without performing any deployments. This change can be achieved safely and easily, without the need to change or release an application's code.

As we will see throughout this book, feature management can be used in several ways within systems to offer new ways of building and developing software. These include the following:

- **Progressive rollouts**: To deliver a new implementation to a production system
- **Experiments**: To gain insight into which variation of a feature performs best
- **Switches**: To use permanent **feature flags** within an application to turn non-essential pieces of functionality on or off
- **Migrations**: To safely move from one system to another

Now, let's look back over the key phases of the DevOps life cycle. Here, I will explain how they can be changed and improved by introducing feature management and its various implementations and use cases.

Revisiting the DevOps life cycle

Let's look at how feature management can be employed within the DevOps life cycle to improve the way software is built and delivered:

- **Planning**: With feature management, smaller amendments to an application can be considered as they can be quickly accessed to understand their value before larger and more expensive work is carried out. Feature management also allows for more risky ideas to be considered since any potential risks to production systems are reduced. All of this allows for an improved **planning** stage.
- **Developing**: Developing new features can be undertaken more safely with feature management as a new implementation can be built alongside an existing one and encapsulated within a feature flag code block (more on this soon). This will allow the feature flag to be used as a toggle and during testing, this can enable a granular approach to discovering where any quality issues might lie. With the ability to switch between old and new implementations, there is less risk of degrading the production system with the work being carried out. It is important to note that this does not mean that quality can be sacrificed to build new, substandard features encapsulated by a feature flag.

- **Delivering**: When in this phase, the value of feature management becomes most apparent as a feature can be turned on or off with a high degree of precision. Crucially, this reduces the risk of deployments to the production systems as features can be deployed but not turned on for users. A release no longer needs to change the customer experience, even though new functionality has been deployed. Now, with feature management, it is up to the team to decide when and to whom the feature should be enabled, with no release required to make this happen. The flexibility here around the scenarios in which a feature can be enabled is what I will be spending time discussing throughout the remaining chapters.

- **Operating**: Even when in the operating phase, feature management adds value as the entire health of the production environment can be improved with the ability to turn features off and on in real time. There are several cases where this can be useful, especially when you can turn performance-intensive features off when experiencing high load or when routing requests through different systems.

Revisiting the four key DevOps metrics

Building on the understanding of how feature management can be used practically, I want to explain how it can positively impact the four key DevOps metrics I mentioned earlier:

- **Deployment Frequency**: Feature management can assist in achieving a better score here as the feature ideas and implementations can now be smaller, making it easier for multiple deployments to happen. In addition, the risk of these deployments is reduced by having the new features turned off when they are released to production.

- **Lead Time For Changes**: Similar to *Deployment Frequency*, a team's performance can be improved by delivering smaller pieces of work and mitigating risk, allowing for quicker releases. There is also the option to bypass some testing environments and processes with feature management, which can save a good deal of time – we will get to this topic in *Chapter 7, Trunk-Based Development*, where I will explain this in more detail.

- **Time To Restore Service**: With the ability to turn features on and off with the flick of a button, feature management enables rapid responses to service incidents. This could be turning off a frontend feature to routing requests to a different backend service.

- **Change Failure Rate**: With feature management, teams can keep the change failure rate very close to 0. As we mentioned previously, if features and changes are deployed to production with the feature flag turned off, then the chance of an issue within production is minimal. Rather, the impact might occur once the feature is turned on, but the blast radius can be far smaller and better managed through targeting which users will be served the feature. Turning a feature off that is causing a problem during an incident can resolve an incident immediately. This allows you to find a real fix for the problem feature to be implemented in a less pressured manner.

From these basic examples, I hope you have a good sense of how feature management can help improve how software can be built and delivered, and some of the ways that its impact can be measured. Throughout this book, we will learn more about the value it can bring to teams and businesses.

Feature management within CI/CD pipelines

With feature management, having its biggest impact within the **delivering** stage of the DevOps life cycle and with the expectation that most people are adopting CI/CD pipelines, I wanted to share a few things before we move on and look at feature management and LaunchDarkly.

I want to make it clear that feature management is not a replacement for good CI/CD processes – it is an enhancement of the CI/CD process that offers more control and opportunities to teams.

For some, the idea of changing features within a production system might seem to go against the merits of a good CI/CD pipeline, in that to get a deployment to the production environment, it must go through several quality gates before it is deemed good enough to be released. But, once in production, that system could be changed, and that leads to the question, *if a release to production could change once deployed, do the quality gates matter so much?* The quality gates do matter and are just as valuable as they would be without feature management. However, the difference is that all possible configurations of the service should be tested within those quality gates before a feature is deemed to be ready for customers. The ability to test all these variations provides the reassurance that whether a feature is *on* or *off*, your service will continue to work as expected.

As we will see throughout this book, feature management is not an end itself, but a means to an end. By implementing feature management within CI/CD processes, new opportunities will present themselves, such as the following:

- **Testing in Production**: This is where a new implementation can be evaluated against a current feature. By using data directly from customers, the new implementation can be proven to be more valuable before it is rolled out to all customers. This covers several topics we will explore, such as rollouts, experimentations, and switches.

- **Trunk-Based Development**: In this case, production can become the only environment for testing, which speeds up development time and reduces the cost of maintaining multiple *production-like* environments.

- **Infrastructure Migrations**: This is where large migrations can be carried out safely and in a methodical manner, even if some parts of a system should be migrated but other parts remain on the existing implementations.

These methodologies and processes can be implemented in other ways but adopting feature management can offer them all within one approach.

Summary

This chapter has given you an introduction to the modern approach to building software and how this can be achieved and measured. We have explored several approaches in which feature management can develop those practices and the tangible ways it can improve CI/CD pipelines and DevOps metrics.

With this knowledge, you should be able to appreciate how feature management can be valuable to the software development process. This, in turn, allows teams to benefit from more efficient practices and from building products with increased value for their customers and their business.

In the next chapter, I will explain how you can build on the DevOps methodology to building software, how you can get the most from feature management, and how you can use LaunchDarkly to make it happen.

Further reading

To learn more about what was covered in this chapter, take a look at the
following resources:

- *State of DevOps Report 2019*, by DORA (DevOps Research & Assessment)
  – available at `https://www.devops-research.com/research.html#reports`

- *What is DevOps?*, by Microsoft – available at `https://azure.microsoft.com/en-gb/overview/what-is-devops/#culture`

2
Overview of Feature Management

The purpose of this chapter is to explain what feature management is, how it empowers software development and product teams, and how it can be used to reduce risk when implementing and deploying new features.

We will look at the two main types of **feature flags**—temporary feature flags and permanent feature flags. There will be a comparison between these types of feature flags with information around some of the use cases that will be explored later in this book, such as **percentage rollouts**, **ring rollouts**, and **switches**. The chapter will then conclude with information about the process of testing in production and the value that this can add to a team, a product, and a business.

In this chapter, we will explore the following topics:

- An introduction to a feature and how it works
- Understanding temporary feature flags
- Understanding permanent feature flags
- Learning testing in production

I will approach these topics by assuming that you have minimal experience with them. By the end of this chapter, you should have a better understanding of what is meant by phased feature management and how it can add value. This will range from simple implementations to the broad concept and approach of testing in production. Additionally, you should gain an appreciation of the merits and uses for both temporary and permanent feature flags. From learning about testing in production, you will know how crucial logging and telemetry are to understanding the impact of offering variations of features to your customers.

With this knowledge, we will be in a good position to begin setting up LaunchDarkly and implementing feature flags.

An introduction to a feature and how it works

Feature management is a simple idea to explain, but its impact and effective implementation within modern solutions can take time. For now, we will just examine what this term means and the types of scenarios it offers before considering the more involved use cases.

Feature management is the name given to the practice of manipulating the experience that is offered within a system without needing to rewrite the code. It is not a new concept, but there are modern approaches to achieving this that empower teams to improve their processes and the products they are delivering.

Feature management is possible through the practice of encapsulating code within an `if` statement and then having a process to determine whether the conditional value is `true` or `false`. Based on that conditional value, a different part of the code is executed; this allows for two or more implementations to be in the code base, but only one is executed per request.

Some of the traditional ways of achieving feature management have included reading a value from a configuration file that could be changed at compile time or through a setting change that would require a runtime restart. However, if we use a more modern approach, specifically, LaunchDarkly's approach, feature management can be achieved through a service that returns the evaluated value based on the context of the request and how the feature flag is configured to target specific users.

This approach to managing features through a service allows for runtime changes within the application without code rewrites, new compiles, deployments, or restarts. As discussed in *Chapter 1, Introduction to Feature Management with LaunchDarkly*, it allows for work to be implemented and deployed but not made immediately accessible to customers. This makes the whole release process less risky, simpler, and quicker, with smaller features being shipped. This is possible because code can be released to production but in a state where the new feature is encapsulated behind a disabled feature flag. Then, at some later stage, the feature flag can be enabled to expose the new functionality to customers.

Before we get into some of the details of feature flags and their implementations, let's consider what is meant by a *feature*.

What is a feature?

I want to make it clear that the idea of a *feature* in the context of feature management or a feature flag does not only refer to a UI component that a customer can interact with. Indeed, a feature really means any functional implementation inside code that can be encapsulated within an `if` statement. As you will see later in the chapter, and throughout the book, the concept of a feature is a broad one, and it is through this wide range of applicable scenarios that feature management becomes such a powerful tool.

Separating releases from deployments

It is important to understand that the ability to manage features offers novel approaches for delivering software. This is because the deployment of new code does not need to equate to a new feature that is being released into an environment. Here, the value is that the risk of a big bang release is removed as, in particular, large and/or risky work can be deployed but not immediately used. The new feature's implementation and performance can be evaluated in a controlled manner before the new functionality is deemed safe and/ or effective enough to be enabled for all customers. We will take a closer look at this type of **rollout** in *Chapter 4, Percentage and Ring Rollouts*.

With deployments no longer resulting in new code being immediately customer-facing, teams can be empowered to deploy features when they choose. Only after certain quality checks have been validated within the production environment does the new feature become available to customers. Note that this separation between a deployment happening independently from when a feature is released is not practical to achieve in traditional ways of delivering software to an environment.

Following is a list that must be considered before deciding when a new feature could be turned on for customers:

- At a specific point in time; for example, to correspond with a company announcement, when other applications are also going live with a new feature, or when a specific event starts or ends.

- Once key stakeholders, customers, or clients have been able to experience the new feature in production and have signed off on it.

- Once the testing (whether manual or automated) confirms that the functionality of the new feature works as required.

- When wanting to gain feedback from customers, who are the *beta testers*, about the new feature.

- When experimenting and rolling out variants to users to gather telemetry regarding their experiences.

- If the product is under significant load or experiencing a production incident, it is possible to carry out the opposite and turn features off to lessen the degradation to customers. This is, sometimes, referred to as **load shedding**.

In *Chapter 3*, *Basics of LaunchDarkly and Feature Management*, we will be discovering how granular **targeting** can be for a feature to enable it for specific users or segments. This granularity makes a number of the preceding scenarios possible. There are many varied reasons and scenarios to warrant turning features on and off. And, while I will explain many of these scenarios throughout this book, you might have your own use cases, too.

Coming back to the topics we explored in *Chapter 1*, *Introduction to Feature Management with LaunchDarkly*, I want to highlight how this methodology does help teams achieve a better **Continuous Integration/Continuous Development (CI/CD)** practice. Now, teams can continuously integrate and deliver new code to the production applications safe in the knowledge that they are not actually changing the customer experience with their deployments.

Explaining how a feature flag works

At this point, I want to detail what a **feature flag** is, as this adds more context to what we will be looking at in the remainder of this chapter. The following is a code snippet from LaunchDarkly that shows you how to implement a feature flag. I will explain the elements within it. I could use several examples that, technically, show a feature flag's implementation. However, for obvious reasons, I am using LaunchDarkly's instead:

```
User user = User.WithKey(username);
bool showFeature = ldClient.BoolVariation("your.feature.key",
user, false);
if (showFeature) {
    // application code to show the feature
}
else {
    // the code to run if the feature is off
}
```

This is the .NET and C# implementation taken from LaunchDarkly's 2021 documentation (the full documentation is available in the *Further reading* section).

As you can see, there is the `if` statement that I mentioned earlier in this chapter, and with the corresponding `else`, we are dealing with two variations of an implementation. However, it is the two lines before the `if` statement that are of real interest here.

The first element I want to explain is the `ldClient.BoolVariation()` function. This is where the call to LaunchDarkly is happening and will return a Boolean value. The first parameter provided, `your.feature.key`, is the name of the feature flag that we are evaluating. We will want multiple feature flags within an application to control different features; naming each flag, and using that name within the code, allows us to encapsulate and control distinct pieces of code and their features. The next parameter is the `User` object, which is instantiated on the first line.

Later in this chapter, I will take a look at `User`. I will also explain it, in greater detail, in the next chapter. `User` can be considered as the context for this feature flag. By this, I mean that we need to provide the context in which we want to evaluate a feature flag to uniquely target each customer (or request) and return the correct value. The final parameter is the default value to be returned by the method. For now, we will not dwell on this, but we will examine it further in *Chapter 3, Basics of LaunchDarkly and Feature Management*.

As you can see, the actual feature flag implementation in code is simple, but it is how this concept is used that really provides the power of feature management. Next, we will take a look at the temporary and permanent approaches to implement feature flags and understand their unique advantages and disadvantages along with some examples of when to use both.

Understanding temporary feature flags

Now that we have looked at feature management and learned how to implement a feature flag, I am going to discuss the use of temporary feature flags. As the name suggests, the point of these feature flags is that they are disposable with a plan to remove them from the code base at some point. Often, this is the best use case for feature management, but there are some scenarios that we will look at later where you will require a permanent feature flag.

The reason feature flags can often be considered short-lived is that flags can be effectively used to deliver new components safely to production. Once a feature is delivered to production, a flag can be used to make the component available to some or all customers. Once the feature has been released for all users, the flag will have served its purpose.

Feature management is often used to better understand how a feature works in production, to become confident with its implementation, and to gain insight into how users interact with it. This offers a better way to manage feature releases within the production systems to enable a better customer experience. Hence, the feature flag is a means to an end: allowing a review and due diligence to take place before customers experience the new feature. Once the feature is proven to add the expected value, or perhaps is proven to do the opposite, then only the successful implementation needs to remain within the code base. The feature flag, including the encapsulation of the unsuccessful variant, can be removed.

However, if there are many flags within the code base, it can become increasingly difficult to manage, especially if there are nested feature flags. There is a second issue, too: with large numbers of feature flags in production, there is a chance of instability within the system, as concurrent features in various states could result in unexpected and untested behavior. Additionally, it can become harder to debug any issues, as it might not be immediately clear which combination of feature flags introduces instability to the production system. For these reasons, we should consider designing temporary features flags.

Next, we will outline some of the ways in which temporary feature flags can be used to help manage releases and reduce any risk to changes within the production environment.

Rollouts

Usually, this scenario is employed when a new feature is expected to objectively deliver better value for the business. This can be via a rearchitecting or more performant implementation of the existing feature or through the addition of a brand new component. This new feature needs to be proven to deliver the expected outcome in production and ensure that no degradation of the system occurs. Once the new implementation is found to be an improvement over the existing one, then the feature will be rolled out to 100% of customers and the existing implementation will be removed in favor of the new feature.

Rollouts, by their very nature, can often be very short-lived, quickly proving the value of a new implementation.

Rollouts—both percentage and ring, where we can enable a feature for an increasing percentage of users or selected groups, respectively, will be covered in greater detail in *Chapter 4, Percentage and Ring Rollouts*.

Experimentations

Experimentations are similar to rollouts but with some crucial differences. Experiments are far more subjective and could result in the conclusion that a new feature is not as valuable to the customers or business as expected. Experiments can be performed for brand new features and to compare new variations against the existing implementation. This use of experiments requires a different mindset. While these **feature flags** are not permanent, they might remain within the code base for a while to ensure that the statistical significance of the results is achieved.

In some cases, rollouts and experimentations might be combined to perform an A/B test. It is prudent to rollout the latest version to a small percentage of customers to ensure that, technically, it works as expected before it reaches the 50% rollout when an effective A/B test can begin.

We will take a far more detailed look at experimentations in *Chapter 5, Experimentation*.

Trunk-based development

Trunk-based development is different from the previous examples we have discussed. In fact, it is a new approach for working with source code and for testing functionality. With this practice, there are no test environments or test deployments involved; rather, all committed work goes straight to the main/master branch (I am assuming Git is the version control system being used) and can be automatically released to production.

Implementations are tested and validated directly in the production environment. Through the use of feature flags, implementations are only enabled by the developers and engineers working on the feature. This approach allows for even greater CI/CD maturity with teams constantly delivering to production even before features are actually completed. Once the feature has been completed, the feature flag can be removed so that the feature is available in production with no encapsulation. This is as long as no other testing or validation is required.

In *Chapter 7, Trunk-Based Development*, we will be taking a far more detailed look at this use of temporary feature flags.

Migrations

Like rollouts, **migrations** are reserved for time-limited and staggered releases of new pieces of functionality. They are often employed when moving from old systems to new systems, allowing for the new apps, resources, or processes to become a part of the main system. Migrations differ from rollouts in a number of ways: their scope can vary due to their size, some parts of the migration might need to happen at different intervals, and various aspects of the migration can be targeted at different groups. In *Chapter 8, Migrations and Testing Your Infrastructure*, we will spend more time on the topic of migrations.

These methods of using temporary feature flags can all be valuable in their own way, but they can be extremely powerful when implemented to enable **testing in production** and the opportunities that they offer. Later in this chapter, we will examine testing in production further.

Managing temporary feature flags can become challenging when the flags have different lifetimes, and some apps might end up with a large number of them. Thankfully, LaunchDarkly provides ways to better understand the state of a temporary feature flag, and in *Chapter 3, Basics of LaunchDarkly and Feature Management*, and *Chapter 9, Feature Flag Management in Depth*, I will be detailing the functionality and information that is available to manage feature flags.

Understanding permanent feature flags

Unsurprisingly, permanent feature flags are the opposite of temporary ones, as they are put into an app with the expectation that they will be there for a long time and are not time-limited. As mentioned in the *Understanding temporary feature flags* section, many flags are often short. However, there are some good reasons and use cases for long-lived flags, too.

Again, it is worth pointing out that applications with a large number of permanent feature flags could be difficult to manage and could easily introduce unexpected instability into a system.

Two common uses of permanent feature flags are **switches** and **entitlements**. Entitlements are a way of enabling or disabling features based on a customer's account tier or subscription. For example, a feature can be served to a VIP customer but not to other customers. In this way, feature management can form part of the business model of a company and be more than just a mechanism to safely release new functionality to a product. We will touch upon this concept later in this book.

I will explain the concept of switches in detail and provide example scenarios in *Chapter 6, Switches*. However, before then, I want to highlight their merits. A switch uses LaunchDarkly to enable or disable functionality at any point across the whole site or for specific groups or customers. This can offer enhanced resiliency to production incidents as some more demanding parts of the system could be disabled to preserve the core functionality. This approach is sometimes described as a **Kill Switch**, **Safety Valve**, or **Circuit Breaker**.

Additionally, this approach can be used where some functionality can be enabled for certain customers or groups. Using feature management in this way can replace more traditional configuration systems. For instance, it could be employed for reoccurring time-sensitive functionality where the flag is automatically changed based on a schedule. Also, switches can be used to turn on debugging processes for specific users and customer journeys.

Learning testing in production

For some, the idea of **testing in production** seems contrary to how testing should be done. For a while, the best practice has been that testing should never be done in production: the impact and cost of finding out that features do not work as expected are far too expensive by the time the feature is already in a customer-facing environment. The potential negative impact to both the bottom line and the time taken to build and deliver a feature makes it a very wasteful approach.

However, when feature management is used, the practice of testing in production removes all that risk. In fact, dynamically enabling features for customers in a production environment does not just make this approach safer, but in several ways, it is actually superior to testing features and functionality on any other environment.

Testing in this manner relies heavily on the use of temporary feature flags, as by their very nature, they are designed to help get new functionality to production before validation is achieved that proves the feature works as designed and/or adds an expected value. The previously outlined use cases—rollouts, experimentations, trunk-based development, and migrations—all fall within the area of testing in production.

Before we go into how testing in production offers such value to teams and companies, I want to spend some time on the traditional and old-school ways of performing testing. Additionally, I want to point out that testing in production is not meant to be seen as a *silver bullet* to testing and that other forms of testing are still required.

Testing on a test environment

Traditionally, to perform testing, it was acknowledged as good practice to have test environments that are *like production* so that the code deployed there should behave just as it would once it is released to customers. In this environment, all the required tests can be performed to demonstrate that the feature works as intended and a sign-off can be given. The problem that this practice has is that it can produce false positives where the test and production environments differ. This discrepancy allows for a feature to reach production but not perform as it did in the test environment. To mitigate this, a lot of time and effort should be spent ensuring that the test environment truly is *like production*.

For more complex systems, the scale at which the test environment needs to operate becomes more significant. Development and testing time is simply spent ensuring that the environment works and is configured correctly. Without good management in place, just as much of a team's time might be spent on maintaining the test environment itself as actually delivering new work—in some cases, it could even require more time. That is before we even get to the scenario where multiple test environments are needed. In larger companies, where a number of concurrent features all need to be tested at the same time, unstable test environments are not workable, so code should be tested on separate, stable environments.

In recent years, the software industry has moved to a predominantly **microservices** architecture, which presents a different take on the traditional test environment: each separate service would require its own test instance. If a microservice test application requires access to another microservice for testing purposes, then ideally, it should be hitting another test environment. However, the test environment of each microservice needs to be maintained by a team as if it were in production so as not to block the testing of other services. Additionally, the team for each microservice might want its own test environment where they can validate any new functionality; they would not want to be exposed to the other microservices. The test environments in a microservice architecture can rapidly get numerous, costly, and hard to manage.

Another element that is worth considering is the ease of access to this test environment. In some cases, it is not easy to perform **User Acceptance Testing (UAT)** on the test environment, especially if the environment is only accessible internally. UAT allows stakeholders and/or customers to experience, feedback, and sign off on new work before it is available to all customers.

There is definite value with a test environment, but it certainly comes with a number of downsides. Let's now take a look at how we can better use our production environment to perform tests.

Production is the best test environment

If a feature passes testing on production, you know it works on production. This might seem unnecessary to state, but it has never been as safe or as easy to test functionality on production as it is now. In the past, we needed other ways to validate functionality ahead of it getting to production in order to safely deliver new features.

Where there is a cost to build, configure, and maintain an environment that is *like production*, this can be removed if the production environment is used for testing in production. This reduction or consolidation of environments alone can save vast amounts of time and money; however, it also opens the door to many other opportunities when production is seen as a safe environment in which testing can and should be carried out. Forrester produced a report on the cost benefits of testing in production in January 2021; you can access the online link in the *Further reading* section.

There are many advantages here, especially when compared to some of the scenarios I have just outlined. In terms of the traditional approaches of keeping an environment to be *like production*, it can be tricky to ensure that all databases are kept in sync, all routing is the same, and all edge cases and journeys are catered for. However, in production, all those things come for free as that is how the system runs.

The crucial element to being able to use the production environment to perform tests is that it must be safe to do so. This is achieved with new features, functionality, and implementations being encapsulated in feature flags and not be turned on for customers at the point of deployment. As mentioned earlier, with approaches such as rollouts, you need to be certain that the deployment will not change the customer experience. Instead, you want each deployment to be unnoticeable, and only through changing the targeting rules of a feature flag should the end user experience change.

With that in place, you are free to target the new feature toward yourself, the testing teams, the key stakeholders, the customers, and more. This manages any risks that a new feature might degrade systems and ensures that the testing is done safely in your production environment.

False positives are removed when taking this testing approach. Initially, a feature will be enabled in production for a small percentage of requests/customers to pass through quality gates, which are either manual or automated. Once validated, the feature will then be enabled for a larger number, even up to 100%. Without a need for new builds or deployments between the test and production environments, there is greater confidence in the test itself. It is no longer the case that testing is completed on a test environment only for an issue to be discovered once the deployment to production is complete.

UAT is easier to achieve, too. Similar to getting a feature through quality gates, a feature flag could be enabled in production for key stakeholders. They have the opportunity to provide their feedback and sign off before it is turned on for a larger group of users. Often, it is much easier for stakeholders to access the production system rather than trying to access a test environment. There is greater trust when testing in production that what stakeholders are experiencing and signing off on is exactly what all customers will experience.

With testing in production, other types of testing and validation options become available that were not possible when using dedicated testing environments. Building on top of the preceding UAT example, it becomes much easier to evaluate features with groups of your most active and engaged customers who can provide qualitative information about features before they are rolled out to all customers. The type of testing where a select group of customers knows they are experiencing early access to a feature is often called **beta testing**. Without being able to test in production, testing features with customers is much harder to achieve and, often, so impractical that it is not done at all. This presents a new opportunity to understand how customers are using your product rather than relying on traditional mechanisms (such as surveys or focus groups) to understand the customer behavior.

However, as I mentioned earlier, this approach is not a *silver bullet* for testing, and you will encounter some scenarios where you will want or need to assess something before it is deployed to production. One common example is with database changes, which are far harder to use feature management on, especially if it involves schema changes. Maintaining a smaller number of test environments can certainly be a more cost-effective way of working. Testing in production can help in this scenario too, as a small number of features would need to be assessed in this manner compared to the more traditional manner with others features reaching production for their sign-off without going through the test environment.

It is worth noting that production is not a pristine environment. Traffic and load could be more unpredictable in production than with a traditional test environment, which could impact the validation of a feature.

Testing in production can be seen as another approach to testing that can be employed alongside more traditional practices, or in some cases, such as with trunk-based development, it goes to the extreme and becomes the only way to test features.

Hypothesis-driven engineering

Perhaps the most significant way in which testing in production goes further than traditional testing is through being able to adopt a **hypothesis-driven engineering** approach to working; this is where quantitative data can be gathered from customers. This allows for a hypothesis for a feature to be confirmed, or not, based on how customers are actually using the product. The customers might not even know what they are experiencing on the product is an **A/B test** or **multi variant test** (MVT), and so their usage of the system is unbiased. Sometimes, with the previously mentioned beta testing, the information that is gathered is not completely impartial, as those customers know they are evaluating something and that can alter their views and answers.

I want to mention MVT here, which often gets covered by the phrase A/B tests but can be more complex due to the fact there are more variants at play than just the two found that are found in an A/B test. Throughout this book, when I use the phrase A/B test, it can be assumed I am referring to MVT too unless stated otherwise.

What makes hypothesis-driven engineering such a valuable approach to the way in which products are built, how teams work, and how companies are run is that before huge amounts of time and effort are put into building features, smaller pieces of work can be implemented to understand whether the customers would even be interested. No longer does a feature need to be built because people think it is what customers want; instead, a feature is built because customers have demonstrated they want it.

In *Chapter 5, Experimentation*, we will take an in-depth look at experimentation. However, before that, I will discuss some of the ways in which hypothesis-driven engineering can be used.

Many of you will be familiar with UI-based A/B tests, and they are a good example of how to run two different implementations at the same time with 50% of customers experiencing one or the other. Telemetry and data are gathered about the UI implementations and the customer experience to reach a point of statistical significance before analyzing the success metrics of the experiment. Then, a decision is made about which implementation actually results in a better outcome for the business, and the successful variation is rolled out to all customers. This could be as simple as a wording or color change through to a completely new page layout.

However, there are many ways in which A/B tests can be made more elaborate, as they do not only need to sit within the client experience. For example, a change could be made within the product that actually results in an increase in value for the business later in the customer's lifetime, which requires a broader view of the experiment being run and the types of data being captured.

Rollouts, which I mentioned earlier, can also comparably benefit from hypothesis-driven engineering. While a new implementation might, technically, be an improvement on what existed before, the value to the business after making the change can be calculated via improved customer engagement. Alternatively, it could be that once the new work has been implemented, despite being technically better, it has actually had a negative impact on the success metrics that the system is measured against. These kinds of scenarios allow for discussions regarding your business values and how teams can make informed decisions about the best course of action.

What hypothesis-driven engineering enables and offers teams and companies is the opportunity to quantify their work with regard to the customer experience and the overall value that it will bring to the business; this mirrors much of what the practice of **DevOps** tries to achieve. With an iterative approach, initiatives and projects can be broken down into smaller pieces of work with insights gained on both performance and effectiveness directly from the customers' usage before large commitments are made to deliver features and products. Targeting key groups or segments of customers is key in this scenario, in order to build something small that proves or disproves the hypothesis.

Where the context (or user) comes in

The key to being able to run effective experiments in production is the context of the request, or to use LaunchDarkly's naming, the **User**. Information contained within this is used to effectively target which customers are part of the A/B test. This targeting allows for an experiment to be performed against a subset of customers and, often, allows the work undertaken to be smaller, as it does not need to cater to all types of customers.

LaunchDarkly's User is an extremely flexible and versatile object, which allows for teams and businesses to provide whatever contextual information they require to perform the necessary targeting. We will be looking at working with the User object later in this book, but it is worth knowing that, essentially, anything can be added to the User object and used as targeting criteria within LaunchDarkly, which makes it such a powerful system.

In *Chapter 5, Experimentation*, and *Chapter 9, Feature Flag Management in Depth*, we will explore, in much more detail, how targeting can be implemented and configured to return different evaluations to our application and alter the production functionality. However, before we get there, it is worth considering some of the common properties that can be used for targeting:

- Authentication state
- Subscription level
- Registered country
- Country of request
- IP address
- Device type (desktop, tablet, or mobile)
- Operating system
- Browser
- Domain

With the ability to target features, based on any combination of the `User` object attributes, it becomes easier to segment your customer base into various groups that can be used to validate hypotheses. This can bring about new insights into your customers' behavior, leading to greater efficiencies and ensuring that work undertaken by the teams delivers on the expected value.

Telemetry, logging, and analysis

As you might have noticed over the last few sections, there is a need to gather a lot of information about the performance of the features and implementations within an experiment to validate the successful variant. I want to point out that, alongside being able to implement feature flags to achieve A/B testing, it is also important to have robust and reliable logging and data collection practices.

Without having reliable data to work with, the value of testing in production is significantly diminished as concrete evidence and decisions cannot be made. Being data-driven is an underlying principle of being able to make the most of the opportunities that testing in production presents. Through knowing the usual state of the system as the *control* (or *baseline*), it is easy to compare the effectiveness and performance of new implementations and features to validate just how well the new variant is performing.

Often, the result of an experiment should be well scrutinized, both when it is successful or unsuccessful, to ensure that the external variables are as expected. For example, during an experiment, there could be an effective marketing campaign that sees higher volumes of traffic to a website, which, in turn, could mean there are different types of customers on the site than usual. Being able to review the logs of the system during the experiment allows for this review and for the results to be trusted. This type of scenario is where production can negatively differ from a dedicated test environment as the production environment is not a controlled or pristine one, so external variables can play a part in the results.

As we will examine later in this book, there is information that LaunchDarkly itself can provide to assist you in viewing your application and feature data. However, you might also want to augment this with your own telemetry to give you a wider view of what is happening to your system while you test in production.

Before we move on from testing in production and hypothesis-driven engineering, I want to make it clear that the effective use of this practice goes beyond just software developers/ engineers and includes the whole technology and product department. It also requires the buy-in of the whole business. Everyone needs to be open to this approach and comfortable with trying out ideas in production that could well prove not to be as effective as expected or even as good as what is currently implemented. However, this approach does allow businesses to quickly identify where they should be spending their time and resources to enable faster progress and gain increased value.

Summary

Throughout this chapter, we looked at what feature management is and what it offers to teams and businesses. We also examined what feature flags are, both temporary and permanent, and the scenarios that both types make available. These sections help us to understand how, in using feature flags, it is possible to use the production environment to do more reliable testing and rely on customers to actually inform the work that is carried out.

Feature management offers new and unique ways to design, build, deploy, release, and test software that extends many of the current ways of delivering software. For modern software companies in today's fast-paced and competitive markets, anything that can be done to ensure that the features being built are of most value to customers is vital. Opportunities such as testing in production and hypothesis-driven engineering are approaches that companies can use to become more effective.

From a technological standpoint, I hope to make clear what is available and how learning about feature management can really empower you, your teams, and your business. With this understanding now in place, next, we will explore how to actually use LaunchDarkly and how it can enable scenarios such as rollouts, switches, and experimentations. In the next chapter, we will explore how to set up LaunchDarkly and implement a working feature flag within an application.

Further reading

- *LaunchDarkly Docs*, by LaunchDarkly. This is available at `https://docs.launchdarkly.com/home`.

- *The Total Economic Impact of LaunchDarkly*, by Forrester. This is available at `https://learn.launchdarkly.com/total-economic-impact/`.

3
Basics of LaunchDarkly and Feature Management

In this chapter, we will build on the knowledge you now have of feature management and provide practical examples of how it can help us build and deliver software in new, more efficient, and safer ways. I will cover feature flag management within LaunchDarkly itself: how to implement a flag within an application, how we can target users with features, how to understand the LaunchDarkly **User** system, and how to perform the necessary targeting.

Before we set up a feature flag, I will walk you through setting up the LaunchDarkly client within an application. This will require us to look at both projects and environments within LaunchDarkly. These will help segregate the feature flags by teams/concerns and by the environments (production, test, and so on) that are used.

To ensure that you have set up LaunchDarkly properly from the beginning, I will also detail the **role management** functionality, which provides permissions and restrictions so that teams work with LaunchDarkly in safe and responsible ways.

In this chapter, we will cover the following topics:

- Getting started with LaunchDarkly
- Understanding LaunchDarkly's **projects** and **environments**
- Learning about **feature flags** and **users**
- Understanding **role management**

In this chapter, there will be code examples and screenshots of LaunchDarkly to allow you to get to grips with the basics of LaunchDarkly and feature management.

By the end of this chapter, you will know how to create a LaunchDarkly account, implement the LaunchDarkly client, and set up a feature flag. We will learn how to target specific users with a flag that will pave the way for learning about more powerful targeting use cases later in this book, such as **rollouts** and **experimentation**. You will also know about LaunchDarkly's **projects**, **environments**, and some of the functionality around role management, which will allow you to start working effectively with LaunchDarkly from day 1.

> **What if LaunchDarkly changes?**
>
> The information (code and screenshots) within this chapter is correct at the time of writing but could well change subsequently as LaunchDarkly progresses and makes continual improvements to its products. To mitigate this, I will be providing links to each sample and screenshot so that you can access it in the future, and make sure you are working with the latest version. I will also be explaining the concepts, so, even if the terminology changes, you will still be able to get to grips with LaunchDarkly and feature management.
>
> The experience I have of LaunchDarkly today might even differ from your own as they could well be running experiments within their own product, resulting in different UIs and functionality being available to segments of their users.
>
> All the links are provided in the *Further reading* section of this chapter.

Technical requirements

Within this chapter, we will be exploring some code samples, and you are encouraged to try these out to implement LaunchDarkly and feature flags. To do that, you will need a computer that will allow you to write and run an application. For the examples I am providing, I am creating them with **Visual Studio (VS) 2019** on a Windows PC and in C# and .NET. The expectation is that you know how to set up a default application within VS. There is no technical reason to use these specific tools and languages and if you rather, you could use Visual Studio Code, a Mac computer, and/or a different language or framework to follow along with the examples.

You can find the code files for this chapter here: `https://github.com/PacktPublishing/Feature-Management-with-LaunchDarkly/tree/main/Chapter%203`. There is both a blank web application template to follow along with and a completed version of the application you can look at, once you've followed all the steps outlined in this chapter.

Getting started with LaunchDarkly

The first thing you are going to need to do to get started with LaunchDarkly is to create an account. Depending on your use case, you could use a 2-week trial with a personal account or you might want to reach out to LaunchDarkly to discuss having a trial account for your company. For this chapter, a free trial personal account will be fine.

You can create an account at `https://launchdarkly.com/packt`.

Once you've created your account, LaunchDarkly will offer you a *quick start tutorial*, which you might find useful as it covers similar topics to this section but without some of the explanations and details I will provide. I will approach things from a different angle and explain some of the steps that LaunchDarkly automatically does for you when your account is created.

To start, we will set up the LaunchDarkly client within your application. LaunchDarkly provides several SDKs, and while you can implement your own system to work with LaunchDarkly via its API, I have yet to encounter a case for this. The supported SDKs are very feature-rich and do several things behind the scenes, such as maintaining sync between the application and the LaunchDarkly service.

LaunchDarkly categorizes their SDKs into client-side, server-side, and mobile SDKs; for most of this book, I will only be looking at the server-side implementations, but where I do refer specifically to the other implementations, I will point this out. The reason for this approach is due to cost implications, with the server-side implementation working out as the cheapest way to work with LaunchDarkly. I will talk more about this in *Chapter 13, Configuration, Settings, and Miscellaneous*, if you want to know why that is the case.

You can find information on implementing the LaunchDarkly SDKs within their documentation at `https://docs.launchdarkly.com/sdk`, but at the time of writing, they support 22 of the main languages and frameworks, including the following:

- .NET
- C/C++
- Node.js
- PHP
- Python
- Ruby

Throughout this book, I will be using the .NET implementation (specifically, .NET 5) when providing walkthroughs and code snippets, but you will find that all the languages are similar and that LaunchDarkly has good documentation for all their supported technologies. It is also worth noting that a lot of this book won't involve code examples; instead, it will focus on what LaunchDarkly offers within the tool itself and the approaches to achieving great feature management. However, to understand how the tool works, it is worth setting up LaunchDarkly in an application. That is what we will do next.

Installing the LaunchDarkly client

I will assume you know how to set up a new ASP.NET web application with Razor Pages, but if not, then you can find an example application in the provided source code at `https://github.com/PacktPublishing/Feature-Management-with-LaunchDarkly/tree/main/Chapter%203/Blank%20Web%20Application` Let's get started:

1. The first thing you will want to do with your brand-new app is to install the LaunchDarkly SDK using the Package Manager with the following command:

```
Install-Package LaunchDarkly.ServerSdk
```

2. Next, the LaunchDarkly client needs to be imported within the startup file or entry point of the application. In the example project, this can be done in the `startup.cs` file using the following command:

```
using LaunchDarkly.Client;
```

3. Finally, the client needs to be instantiated for the entire application to access it and not in a per-request location or middleware. Instantiating the client can be done like so:

```
LdClient ldClient = new LdClient("YOUR_SDK_KEY");
```

However, the client needs to be set up as a singleton as you only want a single instance of the client within your app. It is worth stressing that this needs to be a singleton since, without an implementation like this, the application is likely to experience port exhaustion issues. This occurs because you will end up with multiple LaunchDarkly clients being instantiated, and each one will open many connections to LaunchDarkly. Together, they will eventually consume all the available ports on the server your application runs. In the example project, the singleton implementation for the .NET version of the LaunchDarkly SDK can be found in the `Startup.cs` file like so:

```
services.AddSingleton<ILdClient>(new LdClient("YOUR_SDK_
KEY"));
```

When you created your LaunchDarkly account, an SDK key will have been generated. You can find this in **Account Settings | Projects**. In the preceding example, you need to replace the `YOUR_SDK_KEY` placeholder with the SDK key. You will notice various projects and environments when accessing this SDK key. We will take a look at these terms shortly, but for now, it would be worth using the automatically set up **Default project** and within that, the **Test** environment's SDK key. You can access it here: `https://app.launchdarkly.com/settings/projects`.

Congratulations! You should now be able to run your application and have the LaunchDarkly client initialized. This is the first step in being able to implement a feature flag and while this has been straightforward, this does not allow us to control features within our application.

Before we get into implementing and configuring your first feature flag, I want to explore LaunchDarkly itself to show off some of the concepts that will help you fully understand this tool and ensure you can get the most out of it.

Understanding LaunchDarkly's projects and environments

First, I want to talk about the **projects** and **environments** within LaunchDarkly – I have alluded to these already within this book, and it is worth looking at them now. When you created your account, you might have noticed the top-left of the screen, with text stating **Default Project** and **Production**. These denote the project and environment that you are currently in within LaunchDarkly, respectively. This information is visible at all times when working with LaunchDarkly because it underpins everything you will be doing within LaunchDarkly:

Figure 3.1 – The projects and environment information in the top-left corner of LaunchDarkly

You should create a project and two environments – production and test – by default within LaunchDarkly as this allows us to get going very quickly. It is worth understanding their purpose before we move on and set up our first feature flag.

Projects

Projects are effectively a grouping of feature flags and their associated functionality. They can be used in many ways, with the most obvious and common being for a project for each product team or a specific application to allow controls and restrictions to be put in place. For example, a project could be created for a mobile app and another project for a web app, even when the same team builds both. Further definitions and explanations of projects can be found on LaunchDarkly's website.

It's worth considering that much of your production application could be controlled with feature flags, and that you want to ensure that only the relevant people within your organization can access these flags, so having well-defined scopes for projects is recommended. It is also worth noting that over time, there will be many feature flags, both temporary and permanent, within your applications, and by using projects well, you can help keep the grouping and lists of flags to manageable numbers.

Projects are easy to set up and work with but for now, we will continue to use the default project that has already been set up for us. We will be taking a deeper look at projects in *Chapter 13, Configuration, Settings, and Miscellaneous*, including different approaches to using them.

A project on its own is not all we need, though; before we can create flags, we need environments too.

Environments

I have previously spoken about the various uses of environments within testing software, and LaunchDarkly's environments allow for this process to work well with feature flags. While a project is a way of grouping feature flags and functionality, an environment is used to better configure, test, and implement a feature flag safely before we try it out on the production environment.

This means that when creating a feature flag, it is created in one project and across all environments within that project at the same time. The flag can then be configured independently across the environments to target different users, allowing the code changes to be tested as needed with no risk to production. Then, when the feature is ready for sign-off on the production system, it is easy to have separate targeting rules in the production environment, without this impacting how testing is performed on the other environments.

There must be at least one environment within a project, but there is no upper limit. This enables many testing and sign-off use cases. In my experience, there is usually a need for a testing environment and sometimes a **User Acceptance Testing** (**UAT**) environment, in addition to production.

There are several configuration options available with environments for better management and control within LaunchDarkly, such as colorcoding environments and implementing more restrictions and controls on certain ones. For example, production can have much stricter controls on who can make changes to prevent accidents from happening easily. We will spend more time on environments later in this book in *Chapter 13, Configuration, Settings, and Miscellaneous.*

Much like projects, environments are easy to set up, but for now, we will continue with the two default environments that have been set up for us: **production** and **test**.

Bringing projects and environments together

If you click on the projects and environment information area within LaunchDarkly, you will be shown a list of all the projects that are available, as well as each of the environments for the project that are currently selected. You will notice the colors next to the environment name, and while this might seem trivial, these colors can help us know at a glance if we are in the right environment and whether it is safe for us to be making changes.

Right now, you should be in the **Default Project** area within the production environment. As a best practice, you should select the **Test** environment within the project and environment popup, as shown in the following screenshot. Once in the **Test** environment, the box in the top-left corner should now be orange. It is a best practice to make changes to the test environment first as doing so in production could lead us to misconfiguring the targeting rules and breaking the production application. Having said that, when we get to the topic of experimentation in *Chapter 5, Experimentation*, we will try testing our features directly in production. However, before that, we will work in the **Test** environment of **Default Project** as we learn about LaunchDarkly:

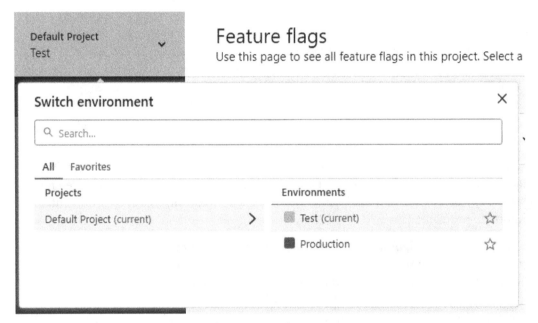

Figure 3.2 – The projects and environment popup

It is important to be aware of the environment you are in when working with feature flags in LaunchDarkly, since making changes in the wrong environment could have a significant impact on your production systems. Knowing which project you're in is less important in this regard as you are unlikely to make changes to a feature flag you didn't mean to, since you should know the correct one's name and/or key.

Now that we have a better appreciation for the projects and environments, we are in a position to create our first feature flag.

Learning about feature flags and users

The first thing we must do when creating and implementing a feature flag is create one within LaunchDarkly. From there, we can implement that flag within the application. You will find the functionality to create a new flag in the **Feature flags** section of the site. Before you create a new flag, it is worth confirming that you are within the **Test** environment of **Default Project**, as outlined in the previous section. You can create your first flag here: `https://app.launchdarkly.com/default/production/features/new`.

When creating a new flag, there are several options and controls available. However, we only need to concern ourselves with the **Name** and **Key** fields for now. I will be covering all the other options and features of a flag throughout the rest of this book, especially in *Chapter 9, Feature Flag Management in Depth*.

The **Name** attribute can be anything you like – it's worth making it meaningful so that it's easy to identify the flag in the future. Following a naming convention for feature flags is highly recommended if you wish to manage large numbers of them in the future. For now, let's call this `My First Feature Flag`. What you should notice is that the **Key** field has auto-populated based on the **Name** field. It should be `my-first-feature-flag`. For now, we will keep the **Key** field as the autogenerated value, but it is possible to set the name of the **Key** field to anything you like. The **Key** field is what is used within your application to identify which flag to evaluate, so it can be good to have a naming convention. Setting a **Name** and **Key** looks like this:

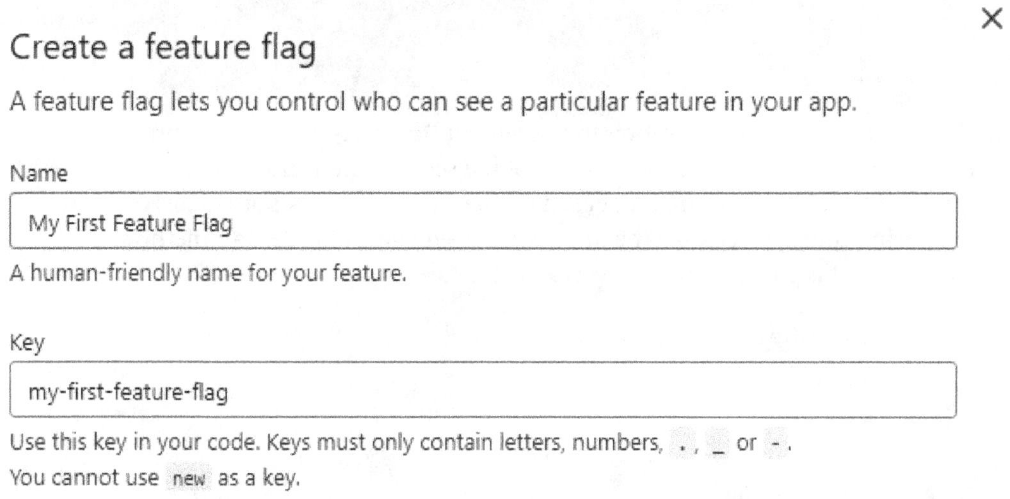

Create a feature flag ✕

A feature flag lets you control who can see a particular feature in your app.

Name

```
My First Feature Flag
```

A human-friendly name for your feature.

Key

```
my-first-feature-flag
```

Use this key in your code. Keys must only contain letters, numbers, `.`, `_` or `-`.
You cannot use `new` as a key.

Figure 3.3 – The Name and Key properties of your first feature flag

Once you click **Save Flag** at the bottom of the screen, LaunchDarkly will now show you the feature flag. There is a wealth of information and functionality on this screen but for now, we don't need to do anything else within LaunchDarkly itself.

We need to go back to our application to implement the flag evaluation and to do that, we need to add the **LaunchDarkly Client (LD Client)** reference within our **Index** page and set up the `User` object to deal with each request on that page.

I won't spend much time on how to add the LD Client to your page or controller as your tech choices might be different, and this will end up not being about LaunchDarkly itself but how different frameworks function. However, I will point out that this step needs to be completed before the rest of the implementation is going to work. The way it is implemented within the example project is shown on the code behind the **Index** page; that is, `Index.cshtml.cs` (this snippet is a simplified version of the whole file):

```
using LaunchDarkly.Client;

public class IndexModel : PageModel
    {
        private ILdClient _ldClient;

        public IndexModel(ILdClient ldClient)
        {
            _ldClient = ldClient;
        }
    }
```

With this in place, we are in a position to implement `User`. As explained in previous chapters, the `User` object is a representation of the session and is used to determine the outcome of a feature flag's evaluation, based on the values of various attributes of `User`. In the following code, you can see the implementation within the `OnGet` method. I will explain what is going on with the setup next:

```
public void OnGet()
{
    User user = LaunchDarkly.Client.User.Builder
        (Guid.NewGuid().ToString())
    .Anonymous(true)
    .Build();
}
```

To create an instance of a user, we need to use the `Builder` method to set the unique Key of the user. In this example, we can just create a new **Globally Unique ID (GUID)** for each request; we will explore more effective ways of doing this later in this book. Next, you will see how we used the `Anonymous` method. From my experience, it is always a best practice to implement this and set it to `true` by default. However, when dealing with logged-in users, you would want this to be `false` as the user would not be anonymous. For logged-in users, there would be some other changes to make to the initialization of the `User` object, but we will get to that later. Finally, we call the `Build` method to create this instance of the `User` object.

At this point, we can implement the evaluation of our first feature flag, since we have set up the LD Client and have access to this request session via the user object. To do this, the code is placed after the user instantiation, as follows:

```
_ldClient.BoolVariation("my-first-feature-flag", user, false);
```

We're using the `BoolVariation` method here as it is the simplest example of a feature flag evaluation, with only a `true` or `false` outcome. It is also the default configuration of a flag, so your first flag to be set up in LaunchDarkly is technically a Boolean one. This method needs three parameters:

- The name of the flag to evaluate. In this case, this is `my-first-feature-flag`.

- The `User` object we initialized.

- The default outcome, should the app not be able to reach LaunchDarkly to perform the evaluation. For this, we will use `false`.

While the third parameter is optional, it is always worth providing it to ensure the correct behavior occurs within your application, should there be any runtime issues. In the *Improving your first feature flag* section, later in this chapter, I will explain what this parameter is used for and why is it advisable to set it to `false`.

Finally, we need to assign this method call to a variable and have that available to the **Index** page itself to show the outcome of the evaluation. The implementation of your first feature flag within `IndexModel` should look something like this:

```
public class IndexModel : PageModel
{

    private ILdClient _ldClient;
    public bool MyFirstFeatureFlagResult = false;


    public IndexModel(ILdClient ldClient)
```

```
    {
        _ldClient = ldClient;
    }

    public void OnGet()
    {
        User user = LaunchDarkly.Client.User.Builder
          (Guid.NewGuid().ToString())
        .Anonymous(true)
        .Build();
        MyFirstFeatureFlagResult =
          _ldClient.BoolVariation
          ("my-first-feature-flag", user, false);
    }
}
```

On the `Index.cshtml` page, I have changed the default HTML to display the feature flag result within the browser. You can copy this into your own project:

```
<div class="text-center">
    <h1 class="display-4">My First Feature Flag</h1>
    <p>Evaluation: @Model.MyFirstFeatureFlagResult</p>
</div>
```

Here, you can see that the `MyFirstFeatureFlagResult` value is being presented within the HTML. After running this code, you should see the following within your browser:

My First Feature Flag

Evaluation: False

Figure 3.4 – The My First Feature Flag evaluation presented within the browser

At this point, we've accomplished a decent amount in terms of setting up our first feature flag and it's a good milestone, but we've not yet proven anything really interesting. Before we celebrate, let's change that **False** to a **True** on the **Index** page. You should keep your application running and go back to the feature flag within LaunchDarkly. At this point, you should see something like the following. You should be able to access it at `https://app.launchdarkly.com/default/test/features/my-first-feature-flag/targeting`:

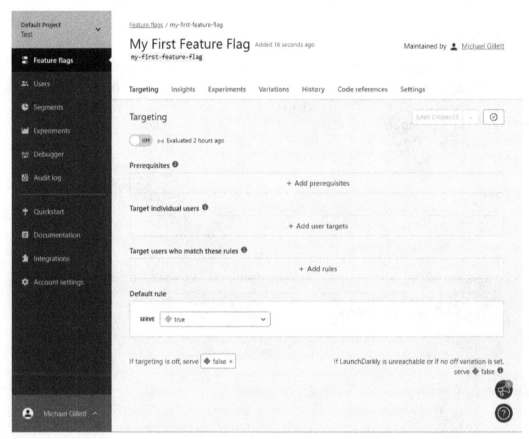

Figure 3.5 – The feature flag details screen in LaunchDarkly

For now, we only need to concern ourselves with the topics of targeting and the flag's default behavior. The first area we should look at is at the bottom of the screen, where it says **If targeting is off, serve false** – this is why we are seeing **False** when running the app as we have not turned on targeting for this flag yet.

A feature flag's targeting is turned off by default and, even once enabled, can be turned off at any point if needed. So, being able to set the outcome of this flag when targeting is disabled is important. By default, a flag is created with a `false` value being returned. Before we turn the targeting on for this flag, we want to know what Boolean value will be returned when enabled to ensure we know what to expect. We can see this value in the **Default rule** area; `true` is the default value for when targeting is turned on.

> **Note**
> It is important to check the setup of the targeting rules before turning on targeting as it will impact all evaluations immediately.

The final thing we need to do here is turn on targeting, which is done via the large toggle toward the top of the page, and then click the **SAVE CHANGES** button. You will be presented with a popup showing an overview of the changes you have made to the flag before you confirm the changes. This might seem overkill for such a simple configuration change, but this is valuable when we get to more complex setups:

Figure 3.6 – Turning on targeting and saving the changes

The next screen will allow you to confirm your changes before you save them:

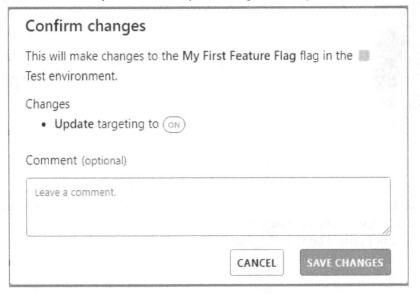

Figure 3.7 – You must confirm your changes before they are saved

Once saved, if you return to your running application and refresh the page, you should see that the evaluation value has changed to **True**:

My First Feature Flag

Evaluation: True

Figure 3.8 – The My First Feature Flag evaluation presented as True within the browser

Now, we have a cause for celebration! You have just implemented and changed the targeting of your first feature flag – congratulations! What you have implemented is effectively a **switch** (since this flag is either on or off) for all users by using the default behavior of having targeting turned on or off. We will now spend much of the rest of this chapter exploring other things we can do with this flag. This will allow you to fully appreciate what is possible with feature management.

Improving your first feature flag

The example we have just gone through should have provided you with an understanding of how to set up a flag, but its implementation isn't quite right for how we plan on using these flags. We will now build upon that example to better represent how we should be implementing flags. This will enable feature management in the scenarios we will explore throughout this book.

As we saw in *Chapter 2, Overview of Feature Management*, what we want to use a feature flag evaluation for is to control whether a piece of functionality is executed or not. We rely on LaunchDarkly to calculate that for us. Our example application does not use the evaluation in this manner; it is only showing us what the flag was evaluated as. In the Index.cshtml file, we want to make a change to bring back the original HTML and only show something substantially different within the browser if the flag is evaluated as true. In our example application, we can replace the markup in the Index.cshtml file with the following code:

```
@{
    if (Model.MyFirstFeatureFlagResult)
    {
        <div class="text-center">
            <h1 class="display-4">My First Feature
                Flag</h1>
            <p>Congratulations! You just managed your
```

```
                    first feature!</p>
        </div>
    }
    else
    {
        <div class="text-center">
            <h1 class="display-4">Welcome</h1>
            <p>Learn about <a
             href="https://docs.microsoft.com/
             aspnet/core">building Web apps with ASP.NET
             Core</a>.</p>
        </div>
    }
}
```

The difference that we now have is that we are using the evaluation from the flag to determine which part of the if statement we want to execute. If you rerun your application now and still have targeting enabled for your first feature flag, you will see a message congratulating you. You should go back into LaunchDarkly and turn off the targeting for the flag and then refresh your application's index page. You will now witness the default text that was originally on the page. We can now toggle between two designs when we run our application:

Figure 3.9 – Left: The application with flag targeting disabled; Right:
The application with the flag enabled

This approach to being able to use feature flags to turn features on and off underpins the entirety of feature management and from this point, we can start to understand how using targeting in different ways can offer us new approaches to building and delivering software. With that, we have expanded our original functionality as we can enable or disable new HTML for all requests (or users, to use the LaunchDarkly naming convention). What we want to be able to do, however, is change the behavior of our application per user and for that, we need to spend some time looking at the User object.

Before taking a look at the `User` object, I want to explain the implementation of the preceding `if` statement. What you will notice is that only with a true evaluation will the new functionality be executed; without a `true` result, the original HTML is presented. The convention is to encapsulate only the new feature code within the successful outcome of `if` and leave the existing implementation within the `else` block.

This convention is advisable when we consider that LaunchDarkly defaults a flag to return `false` when targeting is off. By writing the code in this manner, it creates a default or fallback position for the feature flag that preserves the current implementation. This does not allow new code to be executed without targeting being configured and enabled. You may remember that the third parameter of the `BoolVariation` method was the default evaluation to be returned by the method if LaunchDarkly is unreachable. By having this pattern and providing `false`, you have made your application fault-tolerant to any LaunchDarkly availability issues.

Targeting a user with a flag

In *Chapter 2, Overview of Feature Management*, I discussed the ability to enable features for specific users and segments to gather telemetry, to help validate if an implementation is performing as expected and/or to gather evidence that a feature is something customers are interested in. To effectively achieve this, we need a mechanism so that we can determine whether a flag should be returning `true` or `false` for each user, rather than it being globally enabled or disabled. To implement this, we are going to set up a simple mechanism to alter our request, and therefore the `User` object, on each page load.

This example isn't likely to be a real-world scenario, but it should provide the necessary context to understand the configuration of targeting within LaunchDarkly and how the `User` object can be used. For this, we will introduce a country override value that can be set within a query string parameter. We will pass this through the LaunchDarkly `User` object to allow us to target our new HTML at only those within specific countries. If no country override value is provided, we will set the string to `null` before adding it via the `User` object's `Country` method. To achieve this, our example `OnGet` method looks like this:

```
public void OnGet(string countryOverride)
    {
        string country =
            string.IsNullOrEmpty(countryOverride) ? null
            : countryOverride;
        User user = LaunchDarkly.Client.User.Builder
        (Guid.NewGuid().ToString())
```

```
               .Country(country)
               .Anonymous(true)
               .Build();
        MyFirstFeatureFlagResult =
          _ldClient.BoolVariation("my-first-feature-
        flag", user, false);
     }
```

An optional `countryOverride` parameter can now be provided to our request. If it has a value, then we will use this as our country variable; otherwise, we will treat this as a `null` value. When initializing the `User` object, we will pass this value into the new `Country` method.

Before running our application and testing this change, it is worth going to LaunchDarkly to observe how we can target by country. Within the **Target users who match these rules** section, click to add a new rule. The first box is the attribute you want to be able to target users on, the second box is the type of rule to apply, and the third box will suggest values to use. The fourth box is the outcome that should be returned if a user matches the rule:

Figure 3.10 – How to set up a targeting rule

We will keep this example simple and only use four countries for this example:

- **China**
- **India**
- **UK**
- **US**

We also need to change **Default rule** to `false` so that only those requests with the correct country value will be served `true`. In the following screenshot, you can see how we want the targeting rule to be configured. One thing to check before saving the flag's changes is that targeting is enabled; otherwise, this rule, while being saved, won't be implemented as no evaluations will be performed:

Figure 3.11 – The country targeting rule

When you click **SAVE CHANGES**, you should see the pop-up again, only this time, we should see more useful information than we did previously. As I mentioned previously, the more complex the targeting for a flag, the more useful this dialog window becomes. You should see something similar to the following, showing the list of countries that have been added and the result to be served, and that the default targeting value has been changed:

Confirm changes

This will make changes to the **My First Feature Flag** flag in the Test environment.

Changes

- **Add** rule "If country is one of UK, US, China, India serve ◆ true"
- **Update** default variation to ◆ false

Comment (optional)

Leave a comment.

CANCEL   SAVE CHANGES

Figure 3.12 – Confirm your changes before saving them

Now, we can run our application. By default, you should see the original content load in the browser, but if you append any of the following query string values to the application's URL, you should see the **My First Feature Flag** content on the page:

- `?countryOverride=China`
- `?countryOverride=India`
- `?countryOverride=UK`
- `?countryOverride=US`

To further demonstrate this concept, it is also worth checking what other country override values do. If you set the `?countryOverride` value to any other country, you will see that the original content is returned as those countries don't match any of the ones we configured within LaunchDarkly.

With that, you have successfully implemented a flag and managed which users have been able to access a different feature on the application – this is the basics of feature management with LaunchDarkly.

As I mentioned previously, this example isn't that realistic and it is likely you would want an actual country lookup piece of functionality instead of allowing a user to override their country, but the concept remains the same. You should now understand how targeting a country can be achieved. It is worth considering that LaunchDarkly is not opinionated about the value provided to it for its targeting: what I mean by this is that we used short names for the UK and US. If you were to use the United Kingdom or the United States as the country override values, you would not be shown the same content as when using the UK or the US. Within a real application, it is highly advisable to use an industry-accepted naming convention for your attribute values – I would suggest the ISO country codes in this scenario.

Other attributes to use with the user

I used countries for the previous example as it's a simple concept to grasp but also because it's one of the built-in attributes that LaunchDarkly offers for the user. They have a few other notable ones, including the following:

- Avatar
- Email
- First Name
- IP Address
- Last Name

We don't have an accounting system within our example application, but in systems where you do, I would encourage you to populate as much of this information as possible. We could at least populate the **IP Address** field if we wanted. In my experience, it is often useful to log key user information to LaunchDarkly to easily manage users. Another advantage of providing as much information is that in the future, there might be a need to set up new and more powerful targeting rules. This can be achieved when necessary without needing to deploy changes to the application.

LaunchDarkly does offer a way to include any custom attributes you might require within your applications. Depending on the business and technical needs of your targeting rules, you might find that these custom attributes are used more than the built-in ones. Throughout this book, we will look at examples that can make use of these custom attributes.

The final point I want to make on the subject of attributes is that it is possible to declare attributes as private, which results in the value being used for targeting. However, the data isn't sent to LaunchDarkly and therefore isn't stored within the tool. This allows **Personally Identifiable Information** (**PII**) or confidential information to be used for targeting, but it won't cause issues with data protection regulations. In *Chapter 10, Users and Segments*, we will look at the User object in more detail and cover all these topics.

Users should be unique and consistent

Before we leave this example of setting up your first feature flag, I want to point out that the implementation of the GUID for the user's Key is not a great way of having a unique ID for a user. I wanted to get us up and running quickly, so I didn't address this earlier in this chapter; however, it would be wrong not to spend some time on how we could better approach this.

The idea of the **User Key** is for it to be unique, but consistent, for each user within our application. Our implementation is only offering a unique key, but it is not consistent as it changes every time we refresh the page. When dealing with a system that offers user accounts, you could use any unique value that your product uses, perhaps a username or email address for each account. However, LaunchDarkly's recommendation is to use a unique hash for each user instead.

By having a consistent **Key** value for each user, it becomes much easier to work with targeting and configuring flags to return the required values. It is possible to target flags specifically at individual users, and this can be useful when you're looking to adopt *testing in production* practices as QA engineers, key stakeholders, beta testers. All of this can be set up for the user to experience new features based on their account, regardless of any of the attribute values. With these use cases, we will be dealing with non-anonymous users. In the next part of the example, you will see that we change this part of the application.

Even with anonymous users, there is a need to consistently identify them so that they can receive the same variation of a feature' throughout their time using your product. Without this behavior, they could end up being shown multiple versions of a feature, which would be a poor user experience and would likely distort any metrics being observed to determine the success of a feature rollout.

For our example application, we will use a hash value to identify the user. There are many ways to keep track of the user's hash value – it could be a value saved on the account itself or a runtime-generated value based on the account properties. To keep the examples simple, the following code implements a **GUID** stored in a cookie to maintain a consistent user throughout multiple page loads. While there's several lines of code, most of them are just for cookie management within .NET, so I won't elaborate on everything. This is what the OnGet method in Index.cshtml looks like:

```
public void OnGet(string countryOverride)
{
    string country =
      string.IsNullOrEmpty(countryOverride) ? null
      : countryOverride;

    User user = GetUser(country);
    MyFirstFeatureFlagResult =
    _ldClient.BoolVariation("my-first-feature-
      flag", user, false);
}

private User GetUser(string country)
{
    string ldHashCookieKey = "_ldhash";
    string guid;

    if (!Request.Cookies.ContainsKey
        (ldHashCookieKey))
    {
        guid = Guid.NewGuid().ToString();
        var cookieOptions = new CookieOptions()
        {
            Path = "/",
```

```
            Expires =
                DateTimeOffset.UtcNow.AddDays(365),
            IsEssential = true,
            HttpOnly = false,
            Secure = false,
        };

        Response.Cookies.Append(ldHashCookieKey,
            guid, cookieOptions);
    }
    else
    {
        Request.Cookies.TryGetValue(ldHashCookieKey
            , out guid);
    }

    return LaunchDarkly.Client.User.Builder(guid)
            .Country(country)
            .Anonymous(false)
            .Build();
}
```

The first thing to note here is that I have taken the User instantiation out of the OnGet method itself and moved it to a separate private method (there are other ways to do this but for now, I am keeping this within the same file). Within this new GetUser method, you can see that I looked for a GUID cookie called _ldhash and if that doesn't exist, we generate one and set the GUID to the cookie. When we go to initialize the User object, we will use that generated GUID or read the value from the cookie that already exists. The one final thing to note is that the Anonymous value is now set to false. This is needed as LaunchDarkly doesn't track anonymous users in the same way as known ones. For this example, we are assuming that these are known users to us, even though we don't have an account management system as part of our application. Assuming a user is known is not something that you should do in a real application.

If you run this code now, everything will continue to work as before. However, if you return to LaunchDarkly and navigate to the **Users** page, you should see a single user listed. You can see this at `https://app.launchdarkly.com/default/test/users`:

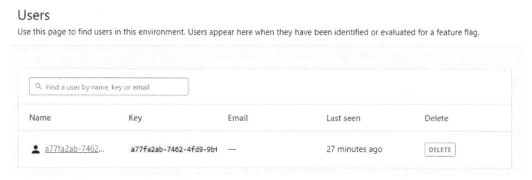

Figure 3.13 – A single user can now be seen in LaunchDarkly

If you click on the user, you can see the information that has been captured by LaunchDarkly and the flags that have been evaluated for this user. If we had set some of the built-in attributes such as name, avatar, and email, we would be able to see those too:

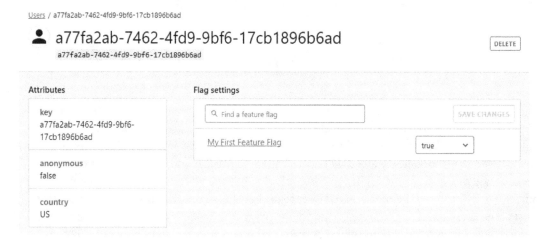

Figure 3.14 – All the information LaunchDarkly has on a user

While there is plenty more to explore with regards to targeting flags and what we can do with users, for now, you should have learned enough that you have a good understanding of how we can achieve feature management, so we will leave the example code at this point. Before we finish this chapter, though, I will explain **role management** and using the correct environments. Remember that so far, we have only been using the **Test** environment.

Understanding role management

To ensure you are on the right path in terms of using LaunchDarkly, the last topic we will cover in this chapter is **role management**. So far, we have been working with the **Test** environment of our **Default Project** as that's the safest way to approach using LaunchDarkly and roles, as well as their associated permissions. This allows us to continue to work with the tool in this manner. We want to restrict who can change flags and targeting rules within the production environment, but other environments are likely to be less restrictive so that our teams can remain productive. Role management can enable these types of scenarios, and many more, through its powerful permission system.

However, before we look at role management, it is worth considering how to work between different environments of a LaunchDarkly project. Right now, our app has the SDK key hardcoded within the code base, which is a bad practice. Instead, we should be using a tool to achieve the following:

- Mask the key so that it can't be recorded and used unintentionally.

- Allow different environments to use different keys.

What tools and processes are available to you are dependent on the technologies you use. Ultimately, we would want the ability to use the **Test** environment SDK key only when deploying to a test environment, and then to use the **Production** key when doing a real release. Taking this further, separate environments could be configured within LaunchDarkly for UAT, beta testing, and so on, and they would generate corresponding keys that would require more test environments to initialize the LD Client against the correct instance of LaunchDarkly.

Once this separation of LD Clients has been achieved, we need to ensure that the separation of these environments is mimicked within LaunchDarkly; otherwise, it would be too easy for production to be accidentally impacted by unintended changes in LaunchDarkly. Remember, you could get to a point where most of the work that's been released to production is deployed behind a flag, so much of the product could be manipulated within LaunchDarkly – with that powerful setup comes the need for good governance.

At this point, the role management topic makes sense when working within teams, so if you are only looking at feature management with LaunchDarkly in an individual capacity, you might get less from this.

A good principle to be followed in the production environment is to grant permission to only those people who need to make changes to the environment. Other members of your LaunchDarkly account can have read-only access to the environment. This setup provides a good level of control over production while allowing the whole team to be able to view and understand the configuration, which can be useful when you need to debug issues.

When I talk about limiting the number of people, there is *no right way* to approach this. In some companies, it could be that only the product team can change production, whereas in other places, developers are empowered to make all the changes; in other instances, QA engineers can control production as they oversee quality. What is right is often up to each team and company based on how they work. I have found that a mix of people across disciplines, and usually those within more senior roles, can offer a good balance of empowerment and risk mitigation.

I will say that as teams become more experienced with feature management, it might be that restrictions for changing production are relaxed to allow teams to move more quickly and deliver a small piece of work, gather insights, and then implement a feature. With a fast-paced work environment and frequent releases of a product, bottlenecks quickly manifest themselves when there are only small numbers of people who are empowered to make changes within production.

There are other controls that can be placed on accounts too, in terms of who is allowed to create new projects, environments, and flags and whether things can be updated. We will look at this in *Chapter 13, Configuration, Settings, and Miscellaneous.*

For now, I will quickly show how a team made up of developers, QAs, and a Product Owner could be set up. The rules for this overly simplified scenario will be as follows:

- Developers will only have access to the test environment as they are not expected to make production changes.

- The Product Owner can only make changes to the production environment.

- QAs will have the ability to change feature flags in both the test and production environments.

You don't need to apply this to your own LaunchDarkly account, but I will provide some screenshots so that you can see how to achieve this yourself. You can create new roles here: `https://app.launchdarkly.com/settings/roles/new`.

Here, you can see that production changes are denied across all the projects for the development team:

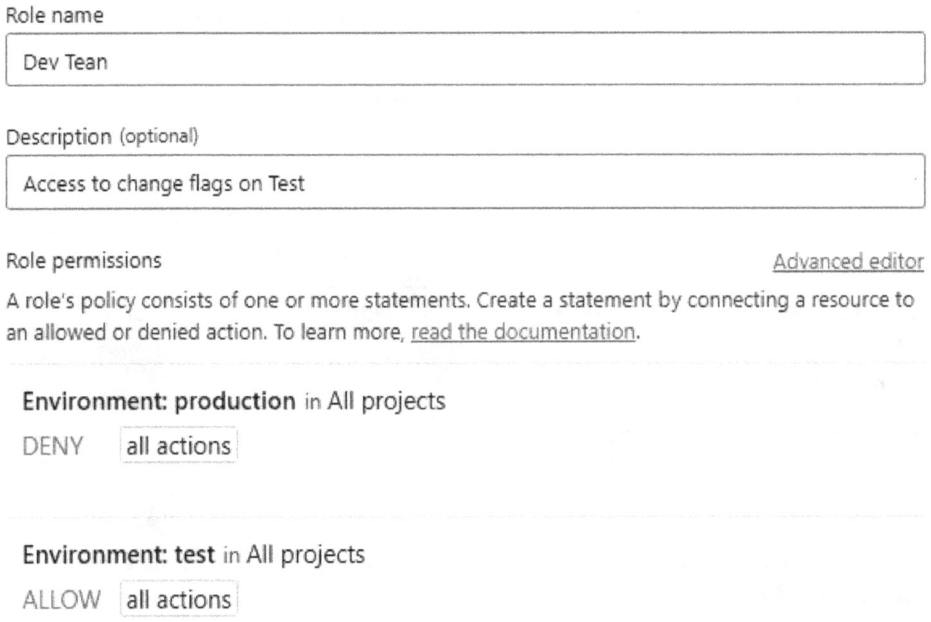

Role name

Dev Tean

Description (optional)

Access to change flags on Test

Role permissions   Advanced editor

A role's policy consists of one or more statements. Create a statement by connecting a resource to
an allowed or denied action. To learn more, read the documentation.

Environment: production in All projects

DENY   all actions

Environment: test in All projects

ALLOW   all actions

Figure 3.15 – The Dev Team's permissions

For the QAs, they can do anything across all the environments within any project:

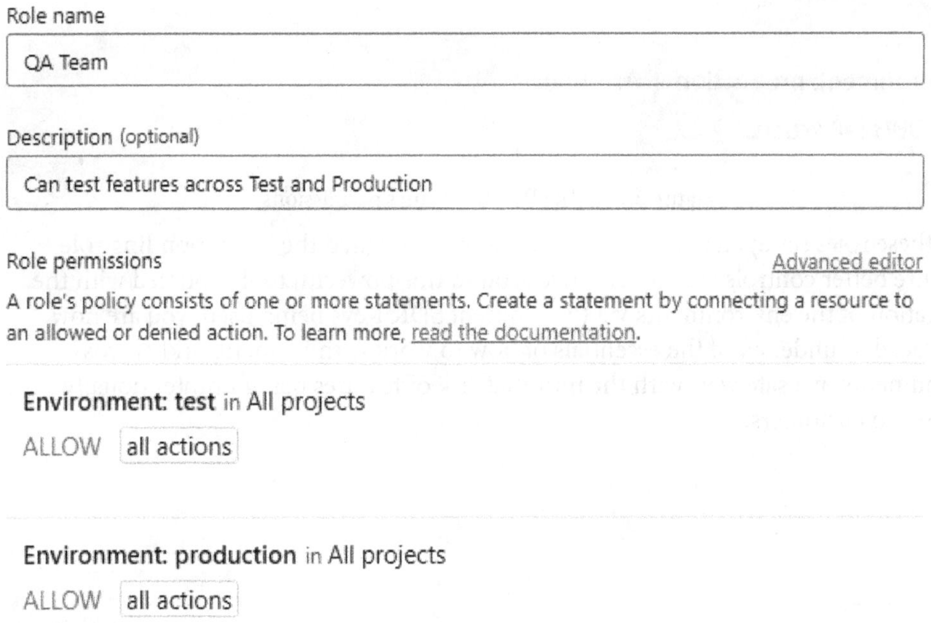

Role name

QA Team

Description (optional)

Can test features across Test and Production

Role permissions   Advanced editor

A role's policy consists of one or more statements. Create a statement by connecting a resource to
an allowed or denied action. To learn more, read the documentation.

Environment: test in All projects

ALLOW   all actions

Environment: production in All projects

ALLOW   all actions

Figure 3.16 – The QA Team's permissions

However, the product team has access to make production changes across all projects:

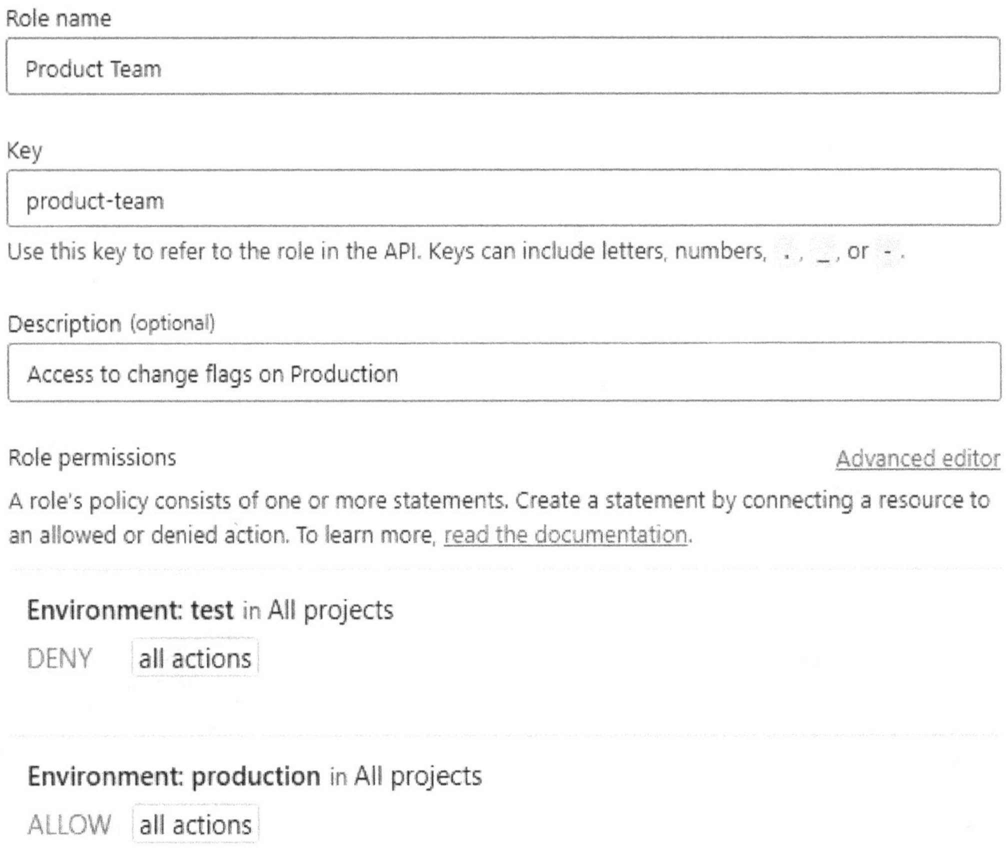

Role name

Product Team

Key

product-team

Use this key to refer to the role in the API. Keys can include letters, numbers, . , _ , or - .

Description (optional)

Access to change flags on Production

Role permissions   Advanced editor

A role's policy consists of one or more statements. Create a statement by connecting a resource to an allowed or denied action. To learn more, read the documentation.

Environment: test in All projects

DENY   all actions

Environment: production in All projects

ALLOW   all actions

Figure 3.17 – The Product Team's permissions

With these roles set up, team members can now be assigned the corresponding role to ensure better controls and governance around this powerful tool. Coupled with the segregation of the environments via the different SDK Keys being used, you are now well placed to understand the essentials of how to work with LaunchDarkly across environments in a safe way, with the minimal risk of features being unintentionally available to customers.

Summary

This chapter has covered the practical ways in which you can use LaunchDarkly, both through the tool itself and how to implement the LaunchDarkly client within an application, which resulted in you setting up and configuring your first feature flag.

Initially, we looked at creating a LaunchDarkly account and how to set up the LD client through .NET code examples. This has given you the skills to do this within any framework and language, and you now have some appreciation for how simple it is to get started with LaunchDarkly.

Next, we explored the topics of projects and environments to make sure you understood these concepts and their purpose beforehand while setting up your first flag. We looked at how to do this within LaunchDarkly and from this, it should be clear to see how easy it is to create flags. With the flag set up, we added it to our example application, where we were able to return `true` or `false` to the application, depending on the state of the flag.

While that example was contrived and not how we would normally implement flags, it did demonstrate how they work. To build on this, we implemented an `if` statement and derived the outcome based on the flag's evaluation, which allowed us to configure the targeting rules. Again, this was not a real-world example but another way in which we can gain a better appreciation for how feature management can be used.

With a new understanding of how targeting works, we explored the `User` object to gain a better insight into what data we can provide. We also changed the `User` object's implementation so that it was more in line with LaunchDarkly's recommendations through a consistent identifier. By providing more data to the `User` object, it will make working with users and targeting easier, and also ensure we can quickly set up new rules in the future without having to make changes to our application.

The final topic I detailed was **role management**. This is not a function of feature management but ensures good governance. It is something you should consider from day 1 to protect your production environment and reduce the risk of unintentional changes being made to your product.

This chapter aimed to provide the basis for a good LaunchDarkly experience and set you on the right path to work effectively with feature management.

This concludes the first section of this book, which has outlined the theory of why feature management is valuable, provided some examples of how it can be used within CI/CD pipelines, and information about the metrics it can positively impact. In addition, this part of this book has provided a practical exercise to show how the theory can be applied both within LaunchDarkly itself and within an application.

Armed with this knowledge, the rest of this book details how feature management is achieved using LaunchDarkly, both in getting the most out of feature management as a practice and how to master LaunchDarkly itself. The next chapter will focus on the percentage and ring rollout use cases for feature management, and you will be shown how to use set up feature flags to achieve both of these types of rollouts.

Further reading

These links are correct at the time of writing. If the links don't work, LaunchDarkly could have changed their routing or the default project or environment names may have changed:

- Sign up for LaunchDarkly: `https://launchdarkly.com/packt`
- View the SDK keys: `https://app.launchdarkly.com/settings/projects`
- Create your first feature flag: `https://app.launchdarkly.com/default/test/features/new`
- My First Feature Flag: `https://app.launchdarkly.com/default/test/features/my-first-feature-flag/targeting`
- View all users: `https://app.launchdarkly.com/default/test/users`
- View role management: `https://app.launchdarkly.com/settings/roles`
- Create a new role: `https://app.launchdarkly.com/settings/roles/new`
- *Getting started with the SDK*, by LaunchDarkly, available at `https://docs.launchdarkly.com/sdk`

Section 2: Getting the Most out of Feature Management

This second section focuses on the uses, scenarios, and opportunities that feature management offers and how LaunchDarkly makes them possible. The section looks at how testing in production can be used to gain feedback directly from customers via percentage and ring rollouts through to full-blown experiments. Additionally, the section looks at switches and migrations in addition to new ways of testing, building, and deploying software with trunk-based development.

This section comprises the following chapters:

- *Chapter 4, Percentage and Ring Rollouts*
- *Chapter 5, Experimentation*
- *Chapter 6, Switches*
- *Chapter 7, Trunk-Based Development*
- *Chapter 8, Migrations and Testing Your Infrastructure*

4
Percentage and Ring Rollouts

In the first few chapters of this book, we mentioned **rollouts**. In this chapter, we will explore what this term means, the value this approach brings to feature management, and how to configure feature flags in LaunchDarkly to perform a feature rollout.

There are two approaches to rolling out a feature in a safe way and validating the value being derived, either as a percentage or as a ring, and in this chapter, we will look at both of these. For these approaches, we need to capture data about the new and existing code we want to release, to allow us to validate that the new functionality is working as expected.

Alongside the more detailed explanations of rollouts will be some example use cases to provide more context for these approaches. Some of these examples are from my own experience, while others will be hypothetical situations that demonstrate the effectiveness of rollouts.

Once the theory and scenarios of rollouts have been covered, we will look at LaunchDarkly to learn how we can use the tool to enable this type of feature management. There will be screenshots to show the steps you will need to take to configure feature flags for both percentage and ring rollouts. You will be able to follow these steps to implement the feature flag from *Chapter 3, Basics of LaunchDarkly and Feature Management*, if you want to try using rollouts. In this chapter, there will be very few code snippets as most of the functionality exists within LaunchDarkly itself.

In this chapter, we will cover the following topics:

- Understanding a rollout
- Using LaunchDarkly for percentage rollouts
- Using LaunchDarkly for ring rollouts
- Combining ring and percentage rollouts

Throughout this chapter, I will assume that you have at least a theoretical understanding of what feature management is. Reading *Chapter 2, Overview of Feature Management,* will provide that context if needed.

By the end of this chapter, you will be able to configure a feature flag to be rolled out to a percentage of users on your application or to defined rings of customers. You will have some understanding of when to use the different types of rollouts and what the trade-offs between them are. You will also know why combining both percentage and ring rollouts can be effective.

Understanding a rollout

Throughout the first three chapters of this book, I have explained how using feature management can help reduce the risk of releasing new features to production, and how new implementations can be proven to be effective before getting released to 100% of customers. Rolling out a feature is how we can achieve this.

There are two types of rollouts: **percentage** and **ring**. You might have heard of this type of feature management being referred to as a progressive rollout, since a feature progresses either through an incremental percentage of customers or through various rings (or groups) of customers. We will look at both of these approaches to understand their value and uses.

One thing to be aware of with rollouts is that they are usually intended to be temporary. The name implies that this approach to feature management is just about getting something **rolled out** and once that is achieved, the feature flag's encapsulation can be removed from the code base. You might not always want to roll out a feature to 100% of your users but only to a predetermined audience. We will look at a few examples later where we might only want to target customers that meet a certain attribute. This will be explored in the *Combining attribute targeting with percentage rollouts* section.

Some of the approaches outlined in this chapter can offer new business models, for example, where a feature can be enabled for certain types or groups of customers. However, I will caution that with those scenarios, once a feature has been rolled out to the intended audience, it would be best to ensure that the configuration of the feature flag is managed differently, such as through a hardcoded configuration within the code base. While LaunchDarkly does offer functionality for this through a permanent feature flag option, it would be unwise to rely on this for your business.

While there are some effective uses of permanent flags, I am only really cautioning against the entitlement scenario. There is nothing wrong with using LaunchDarkly permanently to determine what functionality a certain type of user is entitled to use, but there is a risk. If LaunchDarkly was to ever have availability issues, then some – or all – of your customers might be unable to reach the functionality they are entitled to do so. Either all functionality would be enabled or disabled for customers, which is going to be bad for the business. Having a local system, rather than a remote one, that manages the entitlement functionality is likely to be a safer option when the business model of your product relies on offering several types of users' different functionality. As we explore more about feature management, we will explore other great uses of permanent flags, such as switches, which are detailed in *Chapter 6, Switches*.

Remember that in *Chapter 3, Basics of LaunchDarkly and Feature Management*, we looked at making our app resilient to any availability issues with LaunchDarkly by setting, in code, a default value. If you use LaunchDarkly to configure functionality related to your business model and if any issues occur, this might significantly impact your customer experience. An example of this would be offering different subscription models to your customers and managing the long-term availability through LaunchDarkly, but if the service went down, all customers would experience the code-defined default experience, which is not ideal. Therefore, rollouts should predominantly be considered temporary, especially once a feature has been 100% rolled out.

Percentage rollouts

A percentage rollout is where we enable a feature for an indiscriminate percentage of our customer base. This approach is used to gather usage information about two or more concurrent implementations without any bias within the collected data.

Often, the customers or downstream systems that are experiencing the various features are unaware that they are part of a validation process, so they are unlikely to be able to influence the results. In other words, there is no sign to the user that they are part of a testing exercise, which means that the data gathered, either generated from the telemetry of systems or customers' behavior, is impartial and not skewed.

There are a couple of broad applications for this approach that will be discussed in the following sections.

Validating that a new implementation technically works as expected

By rolling out the new implementation to a percentage of customers – even a small percentage – telemetry can be gathered to prove that the new code is functioning as expected. By using a percentage rollout, the impact to production due to any issues with the new implementation can be well managed before a decision is made to roll out the new feature to all the customers. By changing the percentage of the rollout, the availability of the feature can be progressively increased until the feature is enabled for 100% of customers. This approach even allows for checks to be made to assess how well the feature performs under increased load.

This is certainly a better approach over the traditional way of deploying an application, which would put the new implementation live to all customers and any issues would impact production immediately. This is a good example of how, with feature management, deployments can be decoupled from the release of a new feature.

Examples of this can be anything from a frontend feature that results in users interacting with a new component to a new type of caching system or authentication process. So long as the feature can be encapsulated and performance data can be captured, then a percentage rollout can work well to safely deliver these changes.

Validating that a new feature delivers the expected value

This approach is a cornerstone of **experimentation**, which will be covered in much greater detail in *Chapter 5*, *Experimentation*, but I wanted to outline the scenarios in which a percentage rollout can be used since there are several other opportunities to make effective use of it.

A new feature might have been built but until customers start interacting with it, there will not be evidence that it delivers the value that was expected from it. Using a percentage rollout makes it possible to gain insight indiscriminately from customers about how they use the feature before exposing it to all users. Data needs to be recorded about the feature's usage to understand the value it is bringing, as well as how it is measured against the baseline of the control implementation.

This is a useful exercise to do for nearly all new frontend features as there are often cases where there was high-level confidence that a new feature would perform well with customers, only to find that the users are not interested in it, or worse, it detracts from other valuable features within the application. This approach can be a highly effective way to validate the interest and value of new functionality wherever customer behavior is crucial to understanding the success of a feature. There are scenarios for this kind of rollout on backend systems too.

However, usually, backend systems should provide a consistent experience and set of functionality to the client applications that limits the types of validation that can be achieved. For example, changes to an API response should not be run as a percentage rollout, especially if it includes breaking changes. This is because the client apps might break or build against an API that is changeable. However, changes to cache expiry times could be experimented with in a percentage rollout to prove that the changes bring about the expected value.

Throughout this book, we will look at cases of using percentage rollout in conjunction with other forms of feature management. This is especially relevant when we want to ensure that a new feature technically works before exposing it to a wider group of customers. For example, if we want to run an A/B test with a 50/50 split between two variants (the control and the new feature), we would want to validate that the new implementation's functionality works as expected with a small percentage of customers before exposing 50% of our users to it.

It is worth bearing in mind that with percentage rollouts, each request can be evaluated differently based on the likelihood of them being served `true` or `false`, as set by the configured percentage value. By setting up the `User` object correctly within your application, you can ensure that the customer receives a consistent experience with your system. In *Chapter 3, Basics of LaunchDarkly and Feature Management*, we looked at how to work consistently with the same user across multiple flag evaluations, and the potential variable nature of percentage rollouts is an example of why that consistency is important. Without being able to reliably track the same customer, they may experience a different implementation every time they are exposed to the same feature. For some backend implementations, this might not be an issue, but if this were something within the UI, this could create a less than ideal customer experience.

Ring rollouts

A ring rollout is where there are defined groups of customers, or rings, that a feature can be progressively rolled out to. A feature would move linearly through a sequence of rings. Once a feature has been validated as successful, through measuring the important metrics for it to be shown to be valuable, it would then be enabled for the next ring.

With this approach, specific customers are placed within groups, which results in a feature flag being consistent each time it is evaluated for the users within a group. I will provide some examples of these groups in the following sections. Ring rollouts differ from percentage rollouts since with percentages, a different evaluation could be generated each time for the same user. For ring rollouts, a user will consistently be served the same variation based on the group they belong to.

Sometimes, with ring rollouts, users will be aware that they are part of a validation process, which can result in the collected data being skewed. This could be through both quantitative and qualitative data as customers' behavior can change when they are aware that they are within a certain group, so caution should be used when considering this data.

A few of the scenarios where ring rollouts can be used will be discussed in the following sections.

Alpha, beta, and other testing groups

You could have rings of customers where you want to test the new features to gain their feedback before enabling the feature for all your customers. The idea of being able to alpha and beta test functionalities is not new but, in this manner, features can be deployed once to production and then enabled as needed for the next group.

In this scenario, once enough feedback had been gathered from the alpha testers, whether through telemetry or qualitative feedback, a feature can then be reviewed. If it is providing the expected value and experience, it could move to the beta testers, where the same exercise can be repeated.

These rings, or groups of customers, are usually *opt in* as the customers need to understand they will be exposed to new and often *untested* functionality (this is *untested* in a loose sense as there would still be quality checks, but the feature itself will not have been used by large numbers of customers where edge cases might reveal some bugs). Customers would be aware that features and the whole product might not be as stable or reliable when they are within these rings, and they might need to provide feedback, but they will get to experience new functionality first.

Often, the alpha ring comprises the internal staff of your company, while the beta group consists of the most active and/or vocal customers who you want to hear from as soon as possible. For testing your products, you might even need more groups than just these two. A well-known example of this methodology (although not within feature management, per se) is Google Chrome's Canary, Dev, and Beta channels, which get access to early builds of the browser: `https://dev.chromium.org/getting-involved/dev-channel`.

One thing to consider with this approach to testing is that the users know they are testing features and are aware they get early access, so feedback and telemetry need to be reviewed with this in mind. This group of customers might not be a true reflection of the wider customer base.

A preview user group

To extend the previous example, consider a group of users who want to preview new features – not necessarily to provide feedback to you, but to allow them to experience new features early on for their own purposes. An example of this would be within an enterprise scenario, where an IT department might want to validate how a new feature behaves with its custom extensions and functionality.

While this ring could provide feedback to you about their experience of a new feature, its purpose is more geared toward enabling them to make changes to their own products, should they find any issues. With this approach, you might only have one preview ring for whom you would enable features, with a set duration, before it is enabled for all customers.

Primary and secondary groupings

Within your customer base, you might have groupings of customers that you want to target with new functionality in specific ways. Where this approach can differ from the previously described ones is that this would not be an *opt in* option for customers. It might not even be apparent to the users that they are within a certain group. The reason we consider this as an option is that rolling out a new feature to one of these groups might not be to gather data about the success of a feature; rather, it could be a business decision to provide a new implementation to a group.

Several criteria could be used to determine which group a customer belongs to, based on which they would experience a new feature. These include the following:

- **Customer type**: It is possible to turn on features based on the type of customer, which opens up several testing and business opportunities. For example, you might have VIP customers that you want to offer functionality to before the rest of your customer base. This could form part of a business model around encouraging customers to pay to get early access to features.

  Alternatively, it could be that you would want to test new functionality on your least valued customers first to reduce any potential risk to the revenue stream.

- **User behavior**: By monitoring the interaction of your customers with your products, you might be able to categorize users into certain groups and progressively roll out new experiences based on this. For example, if you have customers who regularly discover and interact with new functionality, then one strategy could be to enable a new feature for them to quickly gather telemetry. In this manner, you should be able to collect the necessary data faster than if the feature were enabled for a percentage of customers, since some would be slower to interact with the new functionality.

  Depending on the number of customers you have, and the data that can be collected about them, you could have a series of groups that the feature can be enabled for before all the customers get the new implementation. This could be employed for rolling out a feature to groups of customers, starting with the most active to the least active, to ensure that the increased load does not have any impact but maximizes the value for the most active users.

- **User attributes**: There could be some features that should be rolled out to groups of users based on their attributes, which could include information such as their country, device type, or network speed. Even with these groupings, you still might want to offer a feature to all customers, but you would want to validate its effectiveness first with a specific subset. An example of how this can be achieved is to test a feature with customers in a certain country first, perhaps due to the specific conditions within that market and therefore how the customers might want to use the product. Once proven to be successful there, it can be rolled out to additional countries before reaching 100% of customers.

These are just the main scenarios for ring rollouts and there will be other uses for this functionality too, based on the needs of the business and products where feature management is being used.

With both types of rollouts, either quantitative or qualitative data is gathered, which helps us validate how the new feature is rolled out either as an increase in percentage or to the next ring. How this data is collected is up to you, but it is important to be able to differentiate between which versions of a feature the customer was exposed to. Knowing whether the user received the A or B variant allows you to understand if the new code provides the expected value. In more scientific phrasing, we could consider the A variant as the control (or baseline) and B as what we are experimenting with.

Next, we will look at how both percentage and ring rollouts can be configured within LaunchDarkly. There won't be code samples for this as it is all about the setup within the tool. However, if you want to try this out yourself, you can use the code sample we worked with in *Chapter 3*, *Basics of LaunchDarkly and Feature Management*.

Using LaunchDarkly for percentage rollouts

To roll out a feature to a percentage of customers, we need to create and configure a feature flag. *Chapter 3*, *Basics of LaunchDarkly and Feature Management*, shows how to set one up. In this example, the only difference you will notice is that the name of the flag is different. For each example, you might want to create a new feature flag. Once you have created a new flag, we will start exploring percentage rollouts by looking at the **Default rule** property of the flag.

When creating a new Boolean flag, the **Default rule** property is set to `true`. In *Chapter 3*, *Basics of LaunchDarkly and Feature Management*, we looked at how this works when the targeting of the feature flag is enabled to serve either `true` or `false`. However, there are three states a flag can be configured to serve:

- **true**
- **false**
- **a percentage rollout**

In this case, we will be looking at the third option, as shown in the following screenshot:

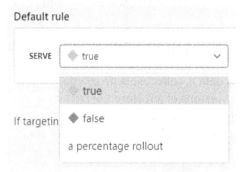

Figure 4.1 – The default rule options of a feature flag

When selecting the **a percentage rollout** option, you will be presented with a new component, where it is possible to configure the percentage of users who should receive the **true** or **false** outcomes.

Within this component, there are two ways to set the percentage for a Boolean flag: either through dragging the slider or typing the percentage value in the inputs. LaunchDarkly ensures you cannot set values to a total greater than 100%. This can be seen in the following screenshot:

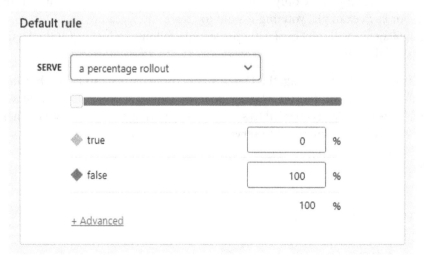

Figure 4.2 – How to set a percentage rollout

You can set the **Default rule** property so that it returns an even 50/50 split, as shown here:

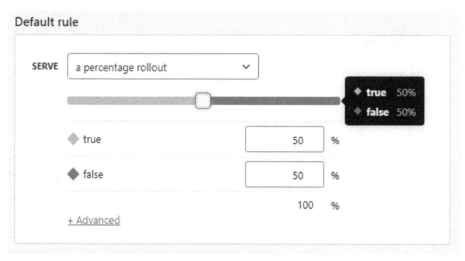

Figure 4.3 – Setting the rollout to 50/50

If you were to test this feature flag following the example given in *Chapter 3, Basics of LaunchDarkly and Feature Management*, you might not immediately see different evaluations being returned to you. This is because we put in place a mechanism to consistently send the same user key (we used a cookie to maintain the same **Globally Unique ID (GUID)** as the key) to ensure that LaunchDarkly could return the same evaluation for each user. It is within this type of scenario that this implementation to consistently serve the same feature flag value makes the most sense, since it ensures that a customer receives a stable experience within your product, even while being a part of the percentage rollout of a feature.

If you want to validate the percentage rollout configuration, all you will need to do is remove the cookie value we were using. Then, there will be a 50/50 chance to get `true` or `false`. It might take a few attempts, but you should see both evaluations for this flag.

Now that you have an understanding of how to set up a percentage rollout, I want to briefly explain the process of doing this on a production system. Normally, it is best to roll out to a small percentage of users, perhaps 5%, to gather enough data to ensure that things are working as expected. Next, the feature would be rolled out to a broader group, maybe 25% or 50%. This can help ensure that not only does the feature work but that it performs well under increasing load. The last step is to take it to 100% for some time. Again, you want to validate that everything is as expected when all the customers are using the feature.

The following screenshot shows how a percentage rollout would usually end up being configured:

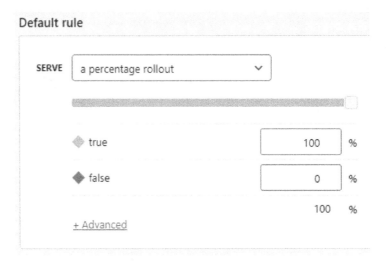

Figure 4.4 – A feature rolled out to 100%

At this point, you should have gained confidence with the new feature after having it run through 100% of the customers. This means that the feature flag can be removed from the code base and the new implementation becomes the only implementation. Once this change has been made and the deployment has been completed, the point of no return has been reached. If any new problems are discovered, then a code change and release would be required as LaunchDarkly can no longer be used to manage the availability of this feature. This is why validating that a feature works as expected becomes crucial throughout the percentage rollout process.

The advanced functionality

You might have spotted the **+ Advanced** label beneath the percentage total of the previously outlined functionality. Opening this submenu presents the ability to change the attribute that is used to calculate the percentage chance of being served `true` or `false` from the feature flag. Any attribute that is added to the `User` object can be selected within this menu.

In most cases, and those I have described earlier, you will want to use the default **key** value. This performs the percentage rollout for each different user as each one has a unique **key**, as seen in the example in *Chapter 3*, *Basics of LaunchDarkly and Feature Management*. If this is changed to **country**, then a feature would be rolled out to a percentage of countries, not to users. Some of the available attributes are shown in the following screenshot:

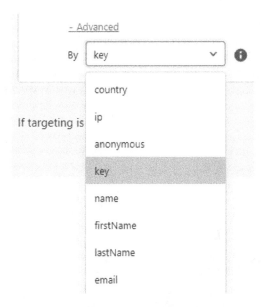

Figure 4.5 – The advanced percentage submenu

By changing the attribute that's used for percentage rollouts, more testing and validation opportunities are made available. However, in my experience, most percentage rollouts are best performed for each user and not the other attributes.

However, LaunchDarkly provides an example where changing the attribute makes sense, and that is within a **Software-as-a-Service (SaaS)** product, where a feature could be rolled out to a percentage of companies rather than users. This results in all users within the same company getting the same experience. In this example, the company name would need to be a custom attribute that is added to the User object when the user is initialized as it is not a built-in attribute.

There are scenarios where you may want to narrow down the sample of customers who might receive a feature via a percentage rollout, since you need some other criteria to have been met before the percentage is evaluated. We will look at how to configure a flag to do this next.

Combining attribute targeting with percentage rollouts

For some percentage rollouts, there is a need to target a subset of your customers and, within that group, enable a feature flag for a percentage of the users. For example, this could be where a feature should be tested for a percentage of users within a certain country.

To achieve this type of rollout, we will follow these steps:

1. First, it is worth setting **Default rule** to `false` and the value to be served for when targeting is disabled to `false`, since we do not want anyone to receive `true` unless they meet our new targeting percentage rule, as shown in the following screenshot:

Figure 4.6 – Setting the default rules to serve false

2. Next, we need to set up a new targeting rule that is only going to be used if the country of the user is **UK**. This is building upon the basic rule we set up in the example in *Chapter 3, Basics of LaunchDarkly and Feature Management.*

3. Then, we need to select the **a percentage rollout** option to serve the user. We will then be presented with the same component we saw previously in *Figure 4.2*. The following screenshot shows how a flag can be configured to target a group of users and, within those users, roll out the feature to a percentage:

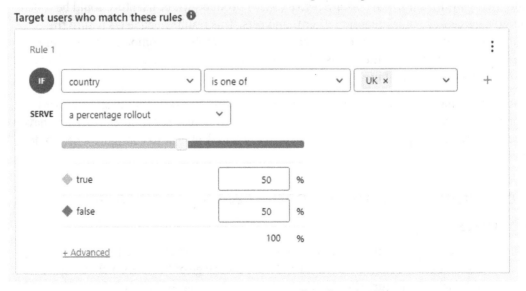

Figure 4.7 – Setting up a percentage rollout for only UK users

It is still possible to change the attribute being used for the rollout. So, in the outlined rule, it could be possible to combine this with LaunchDarkly's example of using the company value, as mentioned in the *The advanced functionality* section. This would result in grouping all UK customers and then only rolling out a feature to a percentage of the companies with UK users.

So far, we have only looked at Boolean feature flags and how they can work with percentage rollouts. But to provide further examples and information about LaunchDarkly, I will detail how the same approach can be used for **multivariate feature flags** in the next section.

Multi-variant percentage rollouts

Before moving on from discussing percentage rollouts, I want to explain how this same technique can be used with multivariate flags too. The flags we have looked at so far have all been Boolean flags, meaning they return a `true` or `false` value, but it is possible to set up a flag that has more than two outcomes. This is known as a **multivariate flag**. While we have not yet looked at this type of feature flag, we shall look at them in more detail in *Chapter 9, Feature Flag Management in Depth*. What follows is a simple explanation of how to set up a multivariate flag.

In *Chapter 2, Overview of Feature Management*, **A/B and multi-variant tests** were discussed regarding **hypothesis-driven engineering**. Boolean feature flags are used for **A/B tests**, whereas **multivariate flags** are used for **multi-variant tests**.

Before showing you how to configure targeting with a percentage rollout, I will show you how to set up a **string multivariate feature flag**. This is done similarly to how we have already created flags. However, in the **Flag variations** dropdown, **String** should be selected, as shown here:

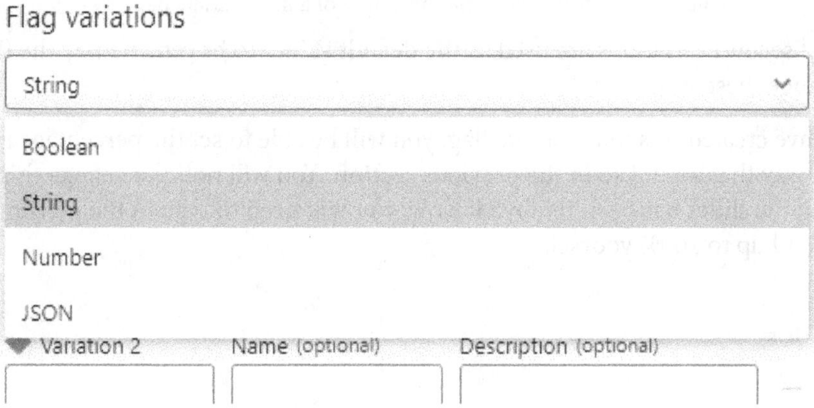

Figure 4.8 – Creating a string multivariate flag

There will now be options for what string values can be served by the feature flag. For this example, I am keeping it simple, and we will have three variations: **A**, **B**, and **C**. You will need to provide the variation value, but the **Name** and **Description** fields can be left blank. Additional variations can be added with the **ADD VARIATION** button:

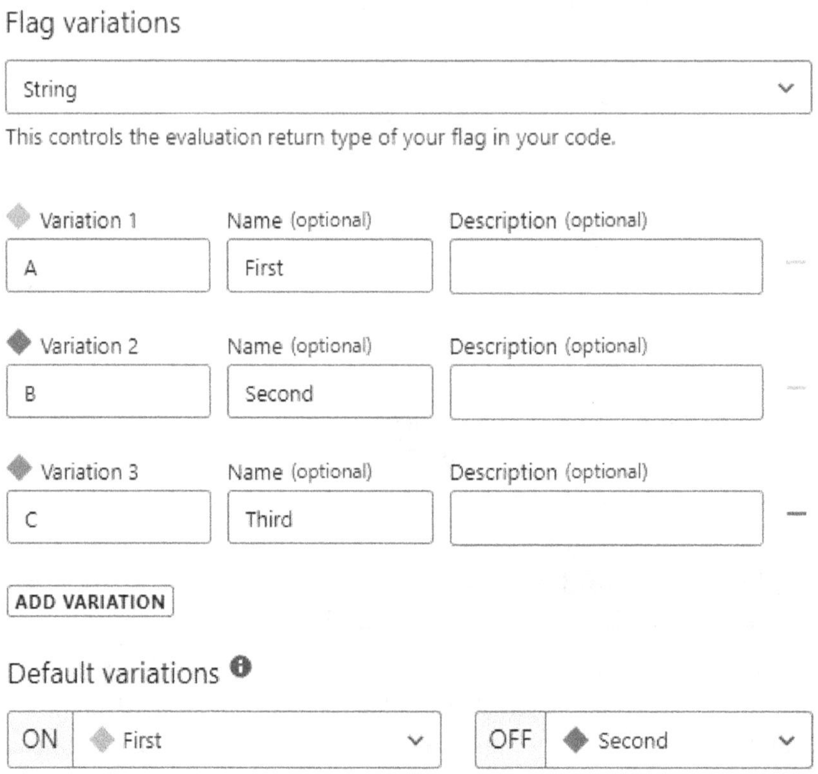

Flag variations

String   ⌄

This controls the evaluation return type of your flag in your code.

◆ Variation 1   Name (optional)   Description (optional)

  A   First

◆ Variation 2   Name (optional)   Description (optional)

  B   Second

◆ Variation 3   Name (optional)   Description (optional)

  C   Third

[ADD VARIATION]

Default variations ❶

| ON | ◆ First   ⌄ | | OFF | ◆ Second   ⌄ |

Figure 4.9 – Setting the variation values of a multivariate flag

The first and second variations are used as the default values to be returned by the flag; you can change these if needed.

Once you have created this multivariate flag, you will be able to set the percentage rollout in the same way that we did so in the previous section. You will notice a change this time, which is that the slider has been removed. Now, you will need to type in the percentage values that add up to 100% yourself:

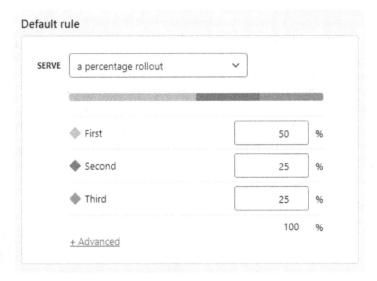

Figure 4.10 – Setting up a percentage rollout for a multivariate flag

As we can see, it is possible to set up the percentages in whatever proportion needed. This can be particularly useful for several scenarios, especially for experimentation, which we will explore in *Chapter 5, Experimentation*.

With rollouts for both Boolean and multivariate flags, it is worth noting how easy it is to set up percentage values and change them. This goes back to the use case of this approach in feature management, where we usually want to roll out to a small percentage of users before expanding the availability by increasing the percentage of customers who will be served `true`. With it being so easy to increase the percentage values within LaunchDarkly, we can quickly release a feature to the necessary number of customers to validate that the implementation works as expected, and then roll it out to all users.

In the next section, we will discover the alternative approach to managing and validating features through rollouts, and that is by using rings, or groups, of customers.

Using LaunchDarkly for ring rollouts

To understand how to use ring rollouts, you will want to create a new feature flag, just as we did for the percentage rollouts. Where this approach will differ is within the targeting. To keep this example simple, I will assume we have three rings to roll out a feature to:

- Alpha testers
- Beta testers
- All customers

In this scenario, we will want a feature to be enabled first for our alpha testers. Then, once the feature is validated within that group, either through quantitative or qualitative data, we would serve `true` for our beta testers and then ultimately make the feature available to all users. To set something like this up, we could use a custom attribute on the `User` object, where we populate a group attribute with the testing group name for all the customers within a group. In this scenario, the code would only add the custom **Group** attribute to the `User` based on some property about the customer's account. There are other ways of grouping customers through the segments system, but we will look at this in the next section.

Using the `User` object code example from *Chapter 3, Basics of LaunchDarkly and Feature Management*, it can be extended to add a group to a user, like this:

```
User user = LaunchDarkly.Client.User.Builder(Guid.NewGuid().
ToString())
        .Custom("Group", "alpha_testers")
        .Build();
```

Then, within LaunchDarkly, we can set up two targeting rules – one for each of our testing groups. Initially, we would only want to target the alpha group with the new feature and have the beta testers and the default value set to `false`:

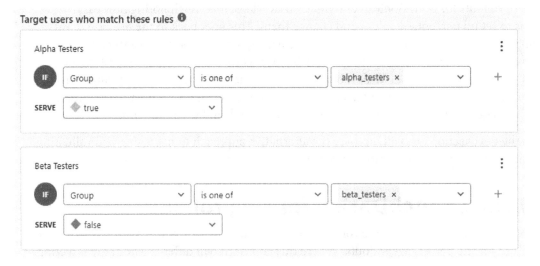

Figure 4.11 – Setting up two testing rings

Over time, feedback can be gathered from the alpha testers, which will provide the confidence or sign-off required before it can be rolled out to the beta testers. This same process can then be followed for beta testers, before a decision is made to roll this feature out to all users. The following screenshot shows how a feature is now enabled for the alpha and beta testers, plus the flag will serve `true` as the **Default rule** property, which will result in the feature being available to all customers:

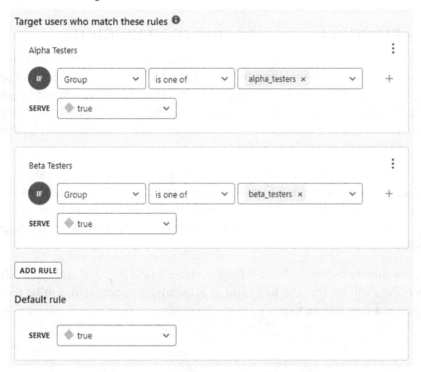

Figure 4.12 – A feature rolled out to all groups

Similar to percentage rollouts, this would be the point where the feature flag encapsulation can be removed from the code base. Once this happens, the new feature would be the only implementation available. At this point, LaunchDarkly will have, once again, assisted in delivering a new feature to your production system.

Again, it is important to collect data to understand whether the feature has achieved the expected functionality and/or value to roll it out to the next ring, and certainly before it becomes the only implementation within the application. How you collect this data and what criteria you would use to ensure the success or failure of the feature is very dependent on the product you are building, as well as the success metric of the feature that has been implemented.

Using segments for easier ring management

Depending on the scenario that you want to use ring rollouts for, you might find that you often target features to your alpha and beta testers, so having to repeatedly set up targeting rules, as seen in *Figure 4.11*, can be a waste of time. Plus, there might be some other mechanisms by which you want to target a user other than the provided examples of the **Group** attribute that were used. To help with this scenario, LaunchDarkly offers segments, which are project-wide groupings of the users.

Taking the previous example of alpha and beta tester groups, we will now create segments for them and see how we would use them within our feature flag for ring rollouts. This becomes particularly valuable if you have a strategy to always use ring rollouts to test features before offering this functionality to all customers. The following are the steps to achieve this:

1. In the **Segments** section of LaunchDarkly, create a new segment and call it **Alpha Testers**. You will see that the **Key** value is automatically populated, similar to when we created a new feature flag:

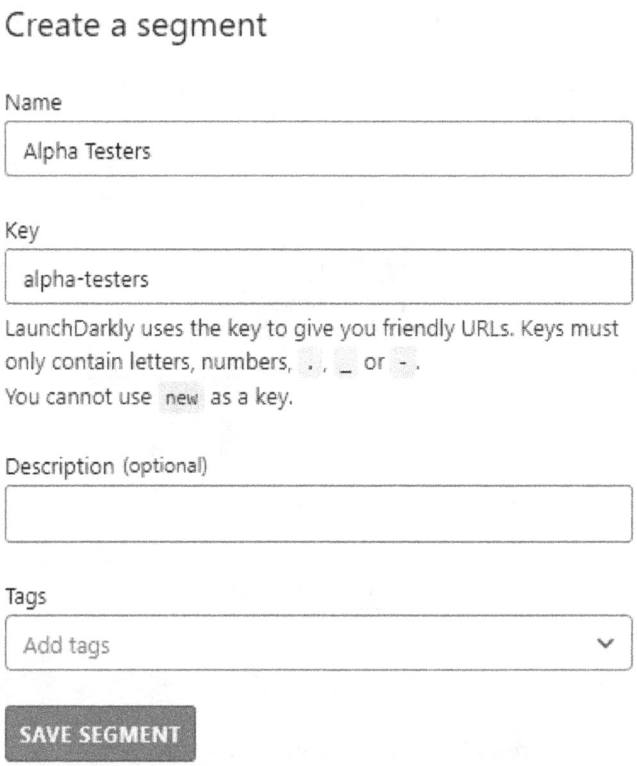

Create a segment

Name

Alpha Testers

Key

alpha-testers

LaunchDarkly uses the key to give you friendly URLs. Keys must only contain letters, numbers, `.`, `_` or `-`.
You cannot use `new` as a key.

Description (optional)

Tags

Add tags

SAVE SEGMENT

Figure 4.13 – Creating a new segment

The segment management page looks similar to the feature flag one, and this is because a segment is an abstraction of the targeting rules that are found within the feature flag page of LaunchDarkly. We will not spend a lot of time here as we will look at segments in greater detail in *Chapter 10, Users and Segments*.

2. For now, we just need to set up a rule to place users within this segment. This will be the same criteria for using the **Group** attribute, which we have used already for users that belong to the **alpha_testers** group:

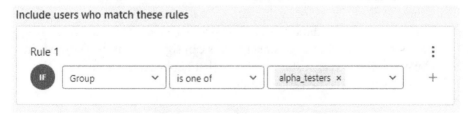

Include users who match these rules

Rule 1

IF   Group   is one of   alpha_testers ×   +

Figure 4.14 – Creating a rule to include users within a segment

3. Now, we can go back to the feature flag and make a change to our targeting rule so that we can use this new segment instead of the targeting rule we had configured. We can also carry out the same process for the **Beta Testers** group to allow us to target a feature at these two segments:

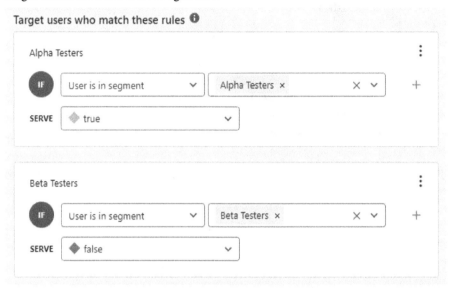

Figure 4.15 – Using segments to target the rings

In our example, it might seem like overkill to create and use segments that just use the **Group** attribute; however, this approach is valuable when you want to target the same rings in the same manner repeatedly across most of the feature flags. It is also useful to be able to create segments of your customer base when using different attributes of their account. For example, there could be a VIP segment for customers with large balances on their account, or a key stakeholders segment that contains individually added users.

Combining ring and percentage rollouts

Before we finish this chapter, I want to discuss of using both ring and percentage rollouts as it provides more opportunities for testing and validating the effectiveness of new features. The most common combination of these two strategies is to start with a ring rollout process, where QA engineers or stakeholders can sign off that the feature is in a position to be presented to real customers. Once that happens, the feature can then be targeted at customers using the percentage rollout.

The following screenshot shows how a feature flag could be configured to target specific users within either a **QA** or **Stakeholders** segment, while also serving all other customers with a 50/50 chance of having the feature enabled. The two segments would have specific users added to them (likely to be people within the company) rather than targeting the feature based on the customer's attributes:

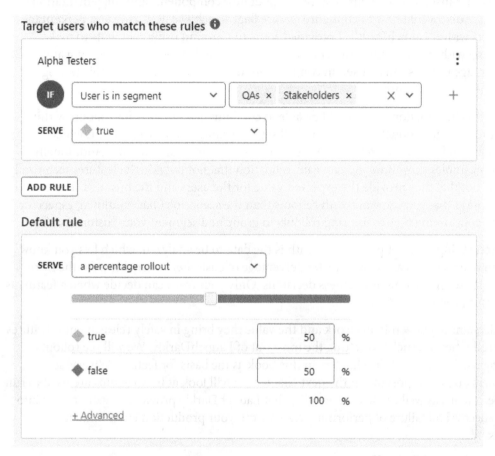

Figure 4.16 – Combining the ring and percentage rollouts

The reason for combining both these approaches is to derive the most value from both ring and percentage rollouts. By using the discriminant method of a ring rollout to target specific users, testing and signing off a feature can be achieved before it is enabled for customers for qualitative testing. The indiscriminate approach to percentage rollouts ensures that the data and feedback that's gathered is unbiased and impartial.

Summary

This chapter should have provided you with an understanding of what is meant by the word **rollout** in the context of feature management. We looked at how progressive rollouts, both **percentage** and **ring**, can offer testing and validation opportunities to help you gather insights from your customers about new components and implementations. Next, we discovered how to configure feature flags to achieve both **ring** and **percentage** rollouts within LaunchDarkly. I provided some tips around using segments to make working with ring rollouts more effective. Finally, I briefly touched on how ring and percentage rollouts can be used in conjunction to offer an even more reliable testing experience for a feature.

With this information, you should be able to configure feature flags for rollouts within your teams and products. I have provided several examples and scenarios in which rollouts can be used to safely deliver new functionality to your products. Additionally, these examples show how you can gain validation that not only do the features technically work, but that they provide the expected value for the users and the business. There will be other ways that you can work with rollouts than the scenarios I have outlined, especially when considering how to use ring rollouts to group and segment your customers.

The one thing I cannot provide you with is the data to be analyzed, which lets you know when to expand a feature to a greater percentage of customers or enable it for the next ring. These will be unique business decisions. Only your team can decide when a feature is ready to progress.

Understanding how rollouts work and the value they bring in safely releasing new features to production is crucial to getting the most out of LaunchDarkly. We will see rollouts being used in many of the chapters in this book as the basis for feature delivery, so knowing this is important. In the next chapter, we will look at how rollouts are used to run experiments, as well as the functionality that LaunchDarkly provides to gain insight into the success and failure of performing tests within your production environment.

Further reading

- To learn more about Google Chrome release channels, please go to `https://dev.chromium.org/getting-involved/dev-channel`.

5
Experimentation

Our focus with feature management so far has been on how to ensure that features can be released to production in a safer and more controlled manner than more traditional approaches. This can be achieved by decoupling the deployment of code from the release of a new feature by making use of rollouts, as we saw in *Chapter 4, Percentage and Ring Rollouts*.

But once it is possible to use feature management within a production application, new opportunities become available for working with new implementations and understanding how users interact with them. In this chapter, we will explore one of the main new opportunities that feature management presents: **experimentation**.

When it comes to how we can use LaunchDarkly to help us experiment, we will be building on our previous discoveries from *Chapter 4, Percentage and Ring Rollouts*, as a rollout is often the best approach to experimenting with new ideas in production. Within that chapter, I spoke about the need to capture data and telemetry to know which variation was the most successful. LaunchDarkly's experimentation tools are one way of capturing and presenting that data.

Before we start looking at what else LaunchDarkly offers for experimentation, I will detail what is meant by the term **experimentation** and why being able to use this approach in production can add so much value to the features you build, the effectiveness of your teams, and therefore the bottom line of your business.

In this chapter, we will explore the following topics:

- What is an experiment?
- Why experiment in production?
- Using LaunchDarkly to experiment

By the end of this chapter, you will gain an understanding of what an experiment is in the context of feature management and LaunchDarkly. You will be shown scenarios in which running experiments on your production application can be of value. We will also explore how to make use of LaunchDarkly to run experiments and view their results. You should then be equipped to run experiments within your applications and start using your customer's feedback, both implicitly and explicitly, to better guide the features that are being built within your product.

It is worth pointing out that LaunchDarkly's experimentation functionality is only provided for enterprise subscriptions. However, the theory part of this chapter does not explicitly rely on the functionality provided by the software; instead, the explanation provided talks mostly in general terms. It is possible to gain insight into the results of an experiment using other tools, but I will only be focusing on how to achieve this using LaunchDarkly within this book.

Technical requirements

When walking through how to set up and configure feature flags and experiments, there will be some code examples and many screenshots of LaunchDarkly. You can use the code samples from *Chapter 3*, *Basics of LaunchDarkly and Feature Management*, to set up an application with LaunchDarkly to test out the LaunchDarkly configuration yourself if you wish.

For the examples I am providing, I am using **Visual Studio 2019** on a Windows PC, C#, and .NET. There is no technical reason to use these specific tools and languages and if you want, you can use Visual Studio Code, a Mac computer, and/or a different language or framework to follow along with the examples.

You can find the code files for this chapter here: `https://github.com/PacktPublishing/Feature-Management-with-LaunchDarkly/tree/main/Chapter%205`. There is a completed version of the application you can look at once all the steps outlined in this chapter have been followed.

What is an experiment?

So far in this book, I have tried to avoid the words *test* or *experiment*, but ultimately, a large part of what we've explored so far has been about running tests and experimenting with new code. When we looked at rollouts in *Chapter 4, Percentage and Ring Rollouts,* I emphasized validating whether the features are technically working and that they are adding the expected value to customers. This can be termed as an experiment.

While I did not state the validation of a **rollout** in the terms of an experiment, there was effectively a hypothesis being validated by rolling out a feature to a **percentage** or **ring** of the customers. For example, when ensuring that something is technically working as expected, the hypothesis would be that the new code is going to work as expected and meet the technical requirements. When we enable that for 10% of the customers and gather telemetry to ensure that this is indeed the case, we are testing the hypothesis. By rolling that feature out to more customers, we are merely expanding the dataset to conclude more confidently that the features are indeed working as needed, before we then release it to all customers.

The other type of validation I mentioned was to rely on customers to validate that a new feature was adding the expected value, whether they are aware they are providing this validation or not. Again, there is a hypothesis that this new feature is going to do something beneficial for our business. A simple example would be changing the text on a **Call To Action** (**CTA**), which may result in more customers engaging with that button. As a result of rolling it out to a percentage of users, their actual interaction with the button with the old or new text shows you which text is more effective.

The previous examples of experiments can be described as proving a **technical hypothesis** or a **business hypothesis**. A **technical hypothesis** is one where a piece of functionality or metric is expected to be achieved, such as improving performance. A **business hypothesis** is one where a feature's success is determined by customers interacting with and making use of the feature as much as possible. In reality, experimentation within the software is not new because usually software is tested before reaching production. This is merely the process of validating the technical hypothesis and checking that what has been built functions correctly. A difference between more traditional testing and the types of experimentation that feature management offers is what the expected outcome will be. When validating that a feature works, there is usually a high degree of confidence that it will perform as expected. This is especially true of technical work. However, when it comes to validating the business value of a feature, an experiment is best utilized when the outcome is unknown.

The value of an experiment is to learn something new. It could conclude that the current feature, not the new one, actually performs better, which is a likely outcome when it comes to testing a business hypothesis. With that in mind, it is worth considering that the smaller an experiment, the better it is for the team and the commercial aspects of the business. This is because if the results of an experiment do not support the hypothesis, then you would want to invest as little time and resources as possible into the new implementation.

An unsuccessful experiment might not mean that the hypothesis itself is wrong, but perhaps the plan to deliver upon it needs refinement. By undertaking small pieces of work to validate the hypothesis, the development teams can quickly iterate on a design (UI, architectural, and/or process) to refine the approach until the expected value is gained from a feature.

I have mentioned the phrase *testing in production* before, and experimentation forms a crucial part of that methodology. In the next section, I will explain why, with experiments, you will want to run them in production. An experiment, or test, does not necessarily have to be exclusively run in the production environment, but it is the environment that is most valuable to you.

Why experiment in production?

The production environment is the best place for you to be running your experiments as you are finding out tangible evidence about your technology and your customers. Other environments can also give you a good indication of how your technology will perform. Good customer research and modeling can provide a reliable indication of what customers want and how they will use the product, but neither is as reliable as actually finding out in production.

In addition to this, there are further opportunities that present themselves when using the production environment for validating technical and business hypotheses, such as **trunk-based development**, which will be explored in *Chapter 7, Trunk-Based Development*.

The kinds of experiments that are likely to be run in production differ from those that would normally be run in other environments. There are two types of features: those that will always form part of the product but need validation and those that might never make it to all customers. It is the latter scenario that really offers new opportunities to hear from our customers and build a better product for them.

Here is an example of using testing in production for a feature that will make it into the production and we have a **technical hypothesis**. Let's say we need this feature to be added to our product and that we do not need to validate that customers interact with it. For example, a new technical implementation is required, and there are some clear criteria that a piece of work must meet for it to function correctly. If this were about improving the performance of a piece of functionality, then there would be an expected level of improvement over the existing functionality (the technical hypothesis). Testing the code on a non-production environment could be done here, but it could be validated in production too via targeting for the testing team, stakeholders, and/or with a rollout to some users. If issues are discovered with the implementation, then they would need to be addressed before the code is released to 100% of customers. This type of functionality will ultimately make its way to production; testing and validation are there to ensure that it performs as expected.

On the other hand, with an experiment that is used to validate a **business hypothesis**, the result of a test could come back as a failure. In this case, this might mean that the feature will never form part of the product. If customers do not engage with the feature, if they do not find it useful, or it does not bring the expected value to the business, then it is not a feature you would want on your product.

For example, if a hypothesis stated that customers would like a brand-new feature that would offer new functionality, several experiments could be performed to prove that was true. These experiments can be run linearly, and each one can inform the next experiment until the whole product is built and tested to ensure a well-refined version of it is shipped. With this strategy, a small piece of work can be undertaken and validated to ensure that the new feature meets everyone's expectations.

Whenever you're working on larger features, I would advise against building and deploying the whole plan since during experimentation, you might find that customers do not engage with it or that it does not provide the expected value. It will have been very wasteful to have invested time and resources into building it, only to find out right at the end that this is not what customers want.

For the example of a large new feature being added to the product, the types of small experiments that could be run to ensure that the feature will deliver the expected value could be as follows:

- **Add a new button for the feature to the site**: From this, you can record the number of users who interact with the button to understand how interested customers are in the particular feature. When a user clicks the button, they can be shown a message about the feature coming soon.

- **Build one user journey**: There are often many journeys for any new feature and by focusing on building one, you can ensure that the simplified experience is something that works for the customers. You can see how many customers go through the whole journey, whether there are points where customers stop engaging with it, and also understand the number of users who want to make use of more functionality and more journeys. You could consider qualitative feedback from customers (especially if you are using a ring rollout) to further understand how customers are finding this feature.

  You might not have a viable product yet, but you do have an understanding of the customer's behavior with this simplified product. The data that's been gathered could reveal that the UI needs to be reworked before most customers will interact with the product, all the way through to discovering that customers are looking for functionality that you hadn't originally planned for.

- **Build the minimum viable product**: You can build out further functionality and journeys to now offer a first truly viable version of the feature or product. There should be a good degree of confidence that it will meet the needs of the customers and the business, but you would still want to experiment as there will be further refinements to make.

With these experiments, you would want to use a **rollout**, either **percentage** or **ring**, as outlined in *Chapter 4, Percentage and Ring Rollouts*. This allows a controlled number of customers to receive these experiments as you only need enough to validate that customers do interact as expected. Once the test has been concluded, the results can be analyzed to understand if work should progress, if refinement of the plan is needed, or if there is no indication that the customers want the feature.

It's worth remembering that through the use of rollouts, at every stage of experimentation (including any others that are identified as useful), the testers and stakeholders can approve what is going to be presented to customers before any customers are exposed to it.

How experiments are managed is quite flexible and each team could have its own process. There are some important things to consider to gain the most from the experiments to ensure they are effective. These include the following:

- **Set a success metric(s)**: Setting which metric(s) will be looked at once the experiment has concluded can help ensure that the data that's collected is useful before any work on the implementation is started. If you don't define a success metric before running an experiment, there is a real risk that, during the analysis phase of the test, doubts about the integrity of the data collected arise, which results in the experiment being inconclusive and a waste of time and resources – the one thing experiments are meant to help with.

- **Know what success and failure look like**: Once you know which metric(s) will be used to gauge the success of the test, it is useful to know what values of it would constitute that the experiment was a success or a failure. The values defined here can help guide whether the experiment was a success, whether the feature needs to be reworked, or whether it should be abandoned entirely.

- **Define the sample size for the experiment**: Before starting the experiment, it is a good practice to agree with all stakeholders on the sample size of customers that will be experimented on. Agreeing on this upfront can ensure there are no misaligned assumptions from stakeholders and development teams about who the experiment will be performed on. There is a range of options that help determine the sample set of customers for an experiment, including users in particular countries, those using specific devices, or through the use of existing segments/rings.

With experiments, it must be accepted by all involved that it might disprove the hypothesis and therefore work should stop. By keeping the work small and the experiments frequent, you can ensure that the team is working on the right features and functionality for the customers. It can feel like this approach results in more work than necessary being done, and this is true in the case of building a fake button to gauge customers' interest. However, the value of knowing that users want a feature far outweighs the cost of building something that does not deliver on the expected value.

There is much more that can be said about experiments as each one is unique. Each feature, team, and business will all have different requirements for the success or failure of an experiment and as you work with this practice, you will discover how valuable testing in production can be.

Next, I will explain how experiments can be set up in LaunchDarkly to provide some more substance to what we have explored so far in this chapter.

Using LaunchDarkly to experiment

When working with experiments within LaunchDarkly, we will still be using feature flags, as we explored in *Chapter 3, Basics of LaunchDarkly and Feature Management*. As we progress through this section, you will want to create some new flags to set up experiments and get to grips with the functionality that LaunchDarkly offers.

The experimentation functionality within LaunchDarkly couples a feature flag with some form of metric that can prove the success or failure of the test. In that regard, we will be using the rollouts from *Chapter 4, Percentage and Ring Rollouts*, to deliver two or more variants of a feature to the users. In the scenario of running an experiment, one of those variants should be the control (baseline or benchmark), which is the current experience that's available to customers. In any experiment, the existing implementation (including when there is not an existing feature) should be considered the baseline. The new implementation that is encapsulated in a feature flag within the code base would be the variant that we then compare the metrics of against the control implementation. This approach can be extended to more than just one variant if where you want to run a **multi-variant test**.

LaunchDarkly can provide us with a visual representation of how well our experiment is performing, showing us how the new implementation does against the control. For LaunchDarkly to show us this experiment information, the first thing we need to do is create a new metric.

You will want to navigate to the **Experiments** section of LaunchDarkly and then select **Metrics**. You will not have any experiments or metrics set up yet. When you create a new metric, you will be presented with a panel, as shown in the following screenshot:

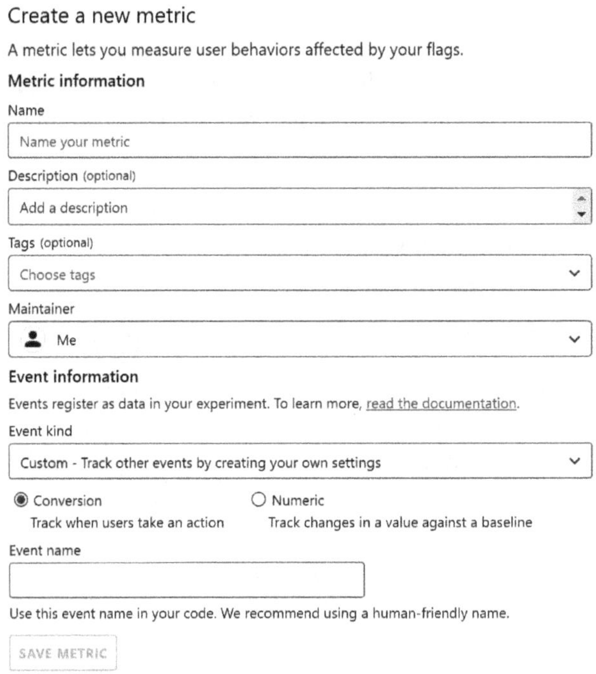

Create a new metric

A metric lets you measure user behaviors affected by your flags.

Metric information

Name

> Name your metric

Description (optional)

> Add a description

Tags (optional)

> Choose tags

Maintainer

> 👤 Me

Event information

Events register as data in your experiment. To learn more, read the documentation.

Event kind

> Custom - Track other events by creating your own settings

⦿ Conversion   ◯ Numeric
Track when users take an action   Track changes in a value against a baseline

Event name

>

Use this event name in your code. We recommend using a human-friendly name.

> SAVE METRIC

Figure 5.1 – Create a new metric page

You will need to provide a name for the metric. For demonstration purposes, I will call the first metric we will work with **CTA clicked**. We will use this metric to record when a button, or CTA, is clicked. You can ignore a number of the fields on this panel for now, but I want to explain the several types of events that can be tracked within a metric.

There are three kinds of events for a metric:

- **Custom**
- **Page load**
- **Click**

For now, we will focus on the **Custom** event type as this is the most flexible and because it can be implemented on the server side rather than in the client. The **Page load** and **Click** events are common events that are used as metrics, so LaunchDarkly offers these as built-in ones. However, they require the JavaScript or React SDKs. You will see this when I explain in *Chapter 13, Configuration, Settings, and Miscellaneous*, that there are some cost-saving aspects to only using LaunchDarkly on the server and not using the client-side SDKs.

Once you have selected the **Custom** event type, you will be presented with two types for the metric, as follows:

- **Conversion**
- **Numeric**

As explained earlier in this chapter, it is worth understanding what it is you are looking to measure in the experiment to know what success and failure look like for the feature you want to test. Knowing this allows you to understand if the outcome you are measuring is a one-time occurrence or if it is a numeric value.

To explain the difference between these two types, I will use a couple of examples. For a one-time occurrence, you can think of this as a registration where the success of the experiment is by getting more users to register. Each user could trigger this **Conversion** event, and it is possible to compare the control and the new implementation in terms of which resulted in more event occurrences. The conversion is expected to be an event that you want your customer to experience.

The other metric type is **Numeric** and is used to record a particular value of a component within your product; for example, the load time of a page. Again, a comparison can be made against the baseline and the new code to understand which variant is performing best. While this value can be recorded at particular events, it might not be an event that the user explicitly performs.

The following screenshot shows how we can set up a **Conversion** metric to capture the number of users who successfully register with our product:

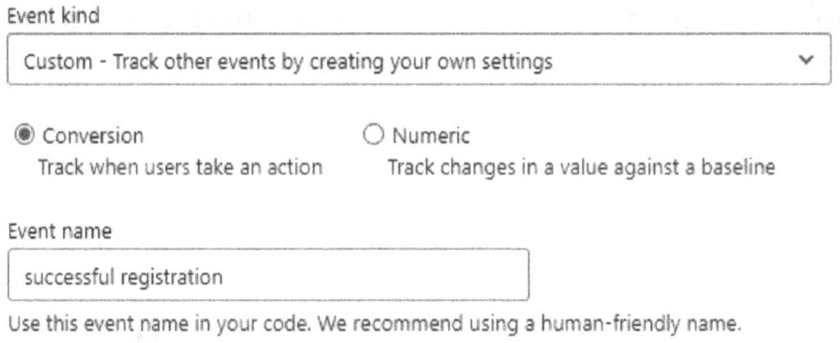

Event kind

Custom - Track other events by creating your own settings   ⌄

◉ Conversion
   Track when users take an action

○ Numeric
   Track changes in a value against a baseline

Event name

successful registration

Use this event name in your code. We recommend using a human-friendly name.

Figure 5.2 – Setting up a conversion custom event

The event name is used within your application's code, within the `Track` method, to determine which metric the user has just experienced. The following snippet shows how the metric in the preceding screenshot can be tracked within the code, It would be added in the code that is executed when a successful registration occurs:

```
_ldClient.Track("successful registration", user);
```

A numeric type of metric is set up in the same way and needs an event name but must also take some additional values. These are the **units of measure** and show which way the value should go against the baseline to prove a successful outcome. To demonstrate this, I have created a second metric called **Load time** and have set up the event like so:

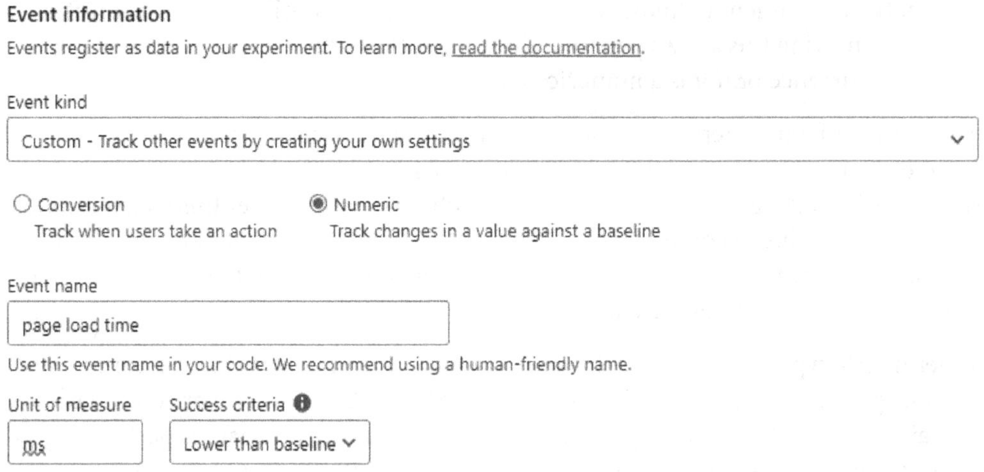

Event information

Events register as data in your experiment. To learn more, read the documentation.

Event kind

Custom - Track other events by creating your own settings   ⌄

○ Conversion
   Track when users take an action

◉ Numeric
   Track changes in a value against a baseline

Event name

page load time

Use this event name in your code. We recommend using a human-friendly name.

Unit of measure   Success criteria ⓘ

ms   Lower than baseline ⌄

Figure 5.3 – Setting up a numeric custom event

To implement a **Numeric** metric in your application, you would again use the `Track` method, but where this differs from a conversion event is that you need to provide the numeric metric value too. The following code sample shows how the metric value could be sent through to `page load event`:

```
_ldClient.Track("page load time", user, LdValue.Null,
metricValue);
```

Once you have both metrics set up, you can view them on the **Metrics** part of the experiments page. The name of the metric and the event that is used within the code can both be edited from this screen if needed.

Next, we will create a feature flag that we can attach to these metrics. This allows us to view what the impact of the flag returning `true` or `false` has on the metrics that we care about. To keep this simple, I am calling this flag **Registration experiment** and it will be a Boolean variation.

On the feature flag, navigate to the **Experiments** screen. Here, you can create a new experiment. First, we want to measure the **Successful Registration** metric with this flag, so select that one from the dropdown list. The baseline for this experiment should be the **false** value. This is because when we encapsulate the feature within our code, we treat **false** as the default. With this approach, we know that the existing code is the baseline. In some scenarios, you might be evaluating a new feature against no feature at all. In that case, you would effectively be encapsulating no code against the new feature's code. You would still want to treat the **false** value as the baseline for your experiment.

The following screenshot shows what setting up a new experiment should look like for our example feature flag and metrics:

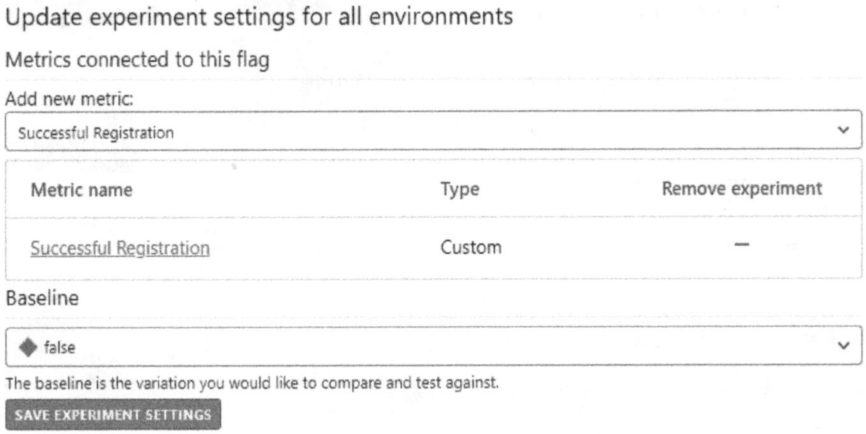

Update experiment settings for all environments

Metrics connected to this flag

Add new metric:

| Successful Registration | ⌄ |

Metric name	Type	Remove experiment
Successful Registration	Custom	—

Baseline

| ◆ false | ⌄ |

The baseline is the variation you would like to compare and test against.

SAVE EXPERIMENT SETTINGS

Figure 5.4 – Adding an experiment to a feature flag

Once this has been set up, you will see that the **Experiment** screen shows some new elements that will show the effectiveness of the feature against our defined metrics. LaunchDarkly will point out that until targeting is enabled, you are unlikely to have significant results, and I would advise against recording any metric results until you enable the flag for the experiment. The recording could be enabled to validate that the correct data is being tracked by LaunchDarkly, but we will want to use a rollout first to get this feature in front of the customers to prove it functions as expected. Once the feature has been proven to work, the experiment can then be run. It would be at this point that the recorded data becomes valuable. If data was recorded before this point, it should be reset before running the actual **A/B test** or **MVT**.

As we saw in *Chapter 4, Percentage and Ring Rollouts*, you might want to use a rollout to ensure that the feature technically works before you even consider running an experiment. In this scenario, you might roll out the feature to 10% of your customers, or to your alpha testers group, to check the telemetry of your application so that you know that everything functions as expected. Then, once you can validate that things work as needed, you can go for a 50/50 split of traffic and start the experiment.

If you rush into running an experiment without validating that both the control and the new implementation technically work, you might end up coming to the wrong conclusions about the feature you are testing. Running an experiment can be done by recording the metrics associated with the experiment. This ability to capture metrics needs to be turned on and can be paused if needed once it is enabled.

The following screenshot shows what an experiment looks like before **Targeting** of the flag is enabled and before any metrics are recorded:

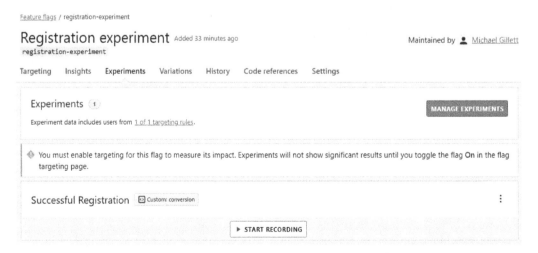

Figure 5.5 – A new experiment set up on a feature flag

To demonstrate a scenario where we can run this experiment, I will have two different pieces of text within a button and we will use the feature flag to determine which text to show. By default, the text will be **Register**, but when the feature flag is enabled for a user, it will be **Join today**. Then, once clicked, the user will go to a successful registration page, which will track the successful registration conversion metric. This scenario emulates how you could experiment with the wordings on your application to discover which words and phrases work best with your customers.

I will detail some of the code we are using for this example and explain what is happening. The `Index.cshtml` page of the example application contains a button link, with the label determined by a LaunchDarkly flag evaluation. The value is rendered from the `@Model.CtaLabel` variable:

```
<div class="text-center">
        <h1 class="display-4">Register today</h1>
        <a href="/success" class="btn btn-primary
          ">@Model.CtaLabel</a>
    </div>
```

Within the code that's executed before the index page is rendered, there is a call to LaunchDarkly to determine which text label to use. The user setup is the same as it was for the index page in *Chapter 3*, *Basics of LaunchDarkly and Feature Management*. The following code snippet shows how this call is implemented and how `CtaLabel` is being assigned a value:

```
CtaLabel = _ldClient.BoolVariation("registration-experiment",
user, false) ? "Join today" : "Register";
```

Next, we need to add a new successful registration page that will live under the path of `/success`, since this is where the button will send the user. The markup of this page is not important but the code for when the page is rendered is important. The rendering of the page is where the successful registration event will be tracked. You can see how this is achieved in the following code block:

```
public void OnGet()
{

    User user = GetUser(null);
    _ldClient.Track("successful registration", user);

}
```

Again, the user is being initialized in the same way as they were in the index page from *Chapter 3, Basics of LaunchDarkly and Feature Management*. I am passing a `null` value in the `GetUser()` method as we do not need the country in this example. Once the code has been implemented, you can run the application and ensure that between having the flags' **Targeting** disabled and enabled, you get two different pieces of text. Then, you can set up the targeting to result in 50% of the users being served **true** or **false**. At this point, you would want to enable the ability to record the **Registration Successful** metric within the experiment screen of the flag.

The following screenshot shows what this experiment looks like once the metrics are being recorded but no events have been received:

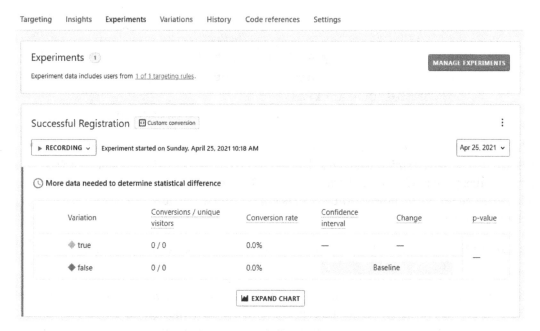

Figure 5.6 – The experiment overview before any data is captured

Now, when you run the app and click the button on the index page to land on the successful registration page, you will see that the event is being tracked within LaunchDarkly. These metric values can take up to 10 minutes to come through to the experiments screen, so don't expect to see real-time usage.

To see the experiment in action, you will want to refresh your application's index page several times and occasionally click the **Register today** or **Join today** button. Remember, if you implemented the cookie logic to consistently use the same user, you will want to clear your cookies to reevaluate the flag. If you click the button more frequently for one of the variant pieces of text, you will see how LaunchDarkly's experiment component can show the successful variant. If you do this enough times, you will reach statistical significance where enough data has been collected for LaunchDarkly to declare a winning variant. However, this is meant to work with large volumes of users and data, so you might not reach this point with your own manual testing by refreshing the page yourself.

The following screenshot shows how the metric values for an experiment are presented once data starts coming through:

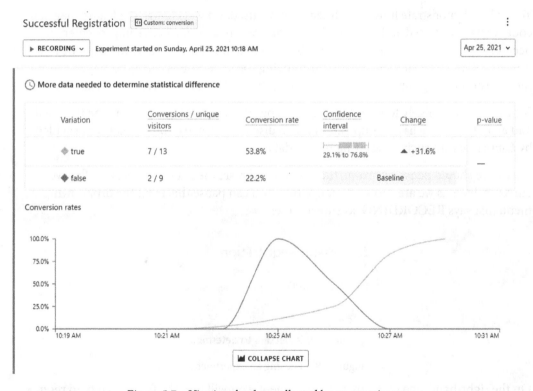

Figure 5.7 – Viewing the data collected by an experiment

With more traffic to an application, the data and graphs become more meaningful than in this simple example. However, what you can see is that when the true value was served, the button was more likely to be clicked on. In this example, this was when the button showed Join today. LaunchDarkly can track these metrics to determine the winning variant. This shows that the winning variant is 31.6% better than the baseline and that from all the data that's been collected, there is a good degree of confidence between 29.1% and 76.8%. The dataset that was used in this example is too small for practical use, so in this case, LaunchDarkly is unable to provide statistical data. In real-world examples with more data, you are unlikely to encounter this scenario.

Next, I want to demonstrate experiments differently so that we can combine conversion and numeric metric events and have two metrics within an experiment.

To further demonstrate how experiments can be used, we will change the feature flag and code we have just used and emulate the scenario where a performance improvement has been made to our application. The experiment would be to understand whether the new implementation is faster and whether improving the load time of the index page results in more customers registering on the site. Again, this will be another contrived example.

In reality, you are likely to want to start fresh, but the approach of tweaking the feature flag and experiments we have set up allows me to discuss a couple of other features provided by LaunchDarkly, such as resetting metric data.

Firstly, we should pause recording the metrics and reset the data in the feature flag we currently have as we are about to change this. You can pause this from the dropdown menu that says **RECORDING**, as shown here:

Figure 5.8 – Pausing an experiment

On the right-hand side of the metric component, you will find the functionality to reset the experiment data, as shown in the following screenshot:

Edit metric
Update the metric used for this experiment in all environments

Reset data
Clear data for this experiment in ▦ Test

Delete experiment across environments
Disconnect the metric and delete experiment data in all environments

Figure 5.9 – Resetting the data we have collected so far

This will only reset the metrics that have been gathered for this environment, which allows implementations of the variants to be trialed as experiments and refined on other environments before being turned on as experiments in production. It also allows for more tests to be done on an environment, without them impacting the same experiment being run on the production environment. It is worth noting that usually, you will not want to change the implementation of a feature flag when you are capturing data, since changes can impact the eventual conclusions that can be drawn.

The next step is to add the **Load time** metric to this feature flag. This can be done in the **Experiments** section of the flag by opening the **Manage experiments** panel. The baseline should be kept as the **false** value. The following screenshot shows how the experiments should be configured for this feature flag:

Metrics connected to this flag

Add new metric:

| Load time | ⌄ |

Metric name	Type	Remove experiment
Load time	Custom	—
Successful Registration	Custom	—

Baseline

| ◆ false | ⌄ |

The baseline is the variation you would like to compare and test against.

SAVE EXPERIMENT SETTINGS

Figure 5.10 – Setting up the experiment for this feature flag

The changes we need to make to our code only affect the index page. We no longer want to change the label of the button; we can now hard code that text to whichever value won the experiment we worked with earlier. From my example, it was Join today. Then, within the code execution on the page being rendered, the feature flag will encapsulate a time delay as the control. For the new implementation, it will render the page as quickly as possible. This is to demonstrate the scenario in which a new, faster implementation is being experimented with. You are unlikely to want to purposely slow down the rendering of a page on your website.

The HTML I am now using on the home page is as follows:

```
<div class="text-center">
        <h1 class="display-4">Register today</h1>
        <a href="/success" class="btn btn-primary">Join
        today</a>
    </div>
```

In the code that is executed to render the page, we will use a stopwatch to record the time it takes to run through the execution. We will provide the elapsed milliseconds of this function as the metric for the numeric event time for the page load time. We will start the stopwatch immediately within the OnGet method and then stop it just before we track the metric against the LaunchDarkly event for page load time:

```
var watch = new Stopwatch();
watch.Start();

string country = string.IsNullOrEmpty(countryOverride) ? null :
countryOverride;

User user = GetUser(country);

if (!_ldClient.BoolVariation("registration-experiment", user,
false))
{
    System.Threading.Thread.Sleep(5000);
}
else
{
    // do nothing
}
```

```
watch.Stop();
double metric = Convert.ToDouble(watch.ElapsedMilliseconds);
_ldClient.Track("page load time", user, LdValue.Null, metric);
```

When you run the application now, you will always be generating a numeric event for the page load time. In those instances, when you click the `Join today` button, you will trigger a conversion event for the successful registration. Again, the dataset will be small here, but even by running this for a few minutes, you will see some interesting elements of the experiment. In a real-world example, we might not know that the new implementation has resulted in reducing the time it takes to load the page, but because we are tracking the time taken, LaunchDarkly knows that the new variant is much faster. When we set the metric up, we selected that a lower value is the better outcome, which enables LaunchDarkly to know how to compare this against the baseline and identify which variant is the winning one for the experiment.

When I ran this experiment, the result was so consistent that even with the relatively small number of attempts that I made, LaunchDarkly was able to reach statistical significance. This is shown by the p-value being less than 0.05 (the threshold value by which statistical significance is achieved) and the green tick mark icon at the top of the metric results. This is expected due to the hardcoded nature of the delay I arbitrarily added to the code. The following screenshot shows what this data looks like:

Figure 5.11 – The results from tracking the numeric event of page load time

In a real-world scenario, you could use this type of setup to prove that a new implementation technically works before trying to prove a **business hypothesis**, where you want to track the users' behavior. For this experiment, we also captured those users that successfully clicked the CTA, and the following screenshot shows the results:

Figure 5.12 – The results of the conversion metric

This shows that when **true** was served and the user had the page load faster, they were more likely to register, which in this example would support the **business hypothesis** that making performance improvements does improve the registration conversion metric.

Being able to combine both user conversion and numeric metrics against a single feature flag allows us to run experiments against our code to ensure things function as we expect, before we even consider understanding the user's behavior.

Summary

From this chapter, you should understand what is meant by the term **experimentation**. Here, we often deal with two distinct types of experiment to validate two different types of hypotheses: **technical hypothesis** and **business hypothesis**.

The **technical hypothesis** is often the first step in delivering a new feature to customers, followed by the **business hypothesis**, which offers innovative ideas to add value to the product and the business itself. We looked at why running experiments in production is the only place to gain the proof needed to support or discredit a **business hypothesis**. Being able to test in production is both safe and valuable when using feature management effectively. Testing in production in this manner poses a negligible risk when deploying code or releasing new features, and a good state can be restored quickly to the production environment if needed, offering a great opportunity to experiment.

In the final section of this chapter, we learned how to use LaunchDarkly itself to measure the success metrics of a test and the tools provided to gain insight into the outcome. We saw how metrics are set against a control, or baseline, which should be the default implementation of code, and how LaunchDarkly uses this information to determine the winning variation. We also learned how both technical and business hypotheses can be validated within a single feature flag by tracking both numeric and conversion metrics.

Experimentation is a valuable aspect of feature management and enables teams and businesses to understand their customers better, and also gain the most value from the features that are being built. By being methodical and iterative with the functionality provided, experimentation allows customers to directly influence the direction the business moves in to best meet their needs – perhaps, all without the customers even knowing they are providing this valuable information.

Throughout the rest of this book, there will be references to **experimentation** and in *Chapter 11*, *Experiments*, we will explore what functionality LaunchDarkly provides in detail. In the next chapter, we will look at using feature management in a new way, where it can be used to introduce **switches** into our applications. This offers the opportunity to turn key pieces of functionality on or off, to help mitigate instability within the product or to gain additional insights.

6
Switches

Switches are a way of working with feature management and LaunchDarkly that we have
not looked at yet. They differ from the likes of **rollouts** and **experimentation** because
they can be a feature flag that is intentionally used permanently within your application.
In the other use cases we have explored so far, the aim has been to release new features in
a controlled and safe manner, which ultimately results in the feature flag being removed
from the code base. However, **switches** offer us a completely different set of scenarios.

Switches can be used to turn key functionality on or off as needed. Toggling functionality
is not done to perform testing or prove that a feature technically works; instead, it is there
to offer teams and businesses the opportunity to respond quickly and easily to changes
within the production environment. Throughout this chapter, we will explore this new
approach to feature management in detail.

In this chapter, we will be covering the following topics:

- Exploring switches
- Discovering use cases for switches
- Learning to implement switches within LaunchDarkly

By the end of this chapter, you will have an understanding of what is meant by the term
switch and the two ways in which they can be used. You will gain knowledge of the
types of scenarios where using switches can add value to your application, as well as the
cases where you shouldn't use permanent feature flags. After that, you will learn how to
implement switches within your application using LaunchDarkly.

As you will discover in this chapter, there is no real difference in the implementation of a feature flag for our previous examples or switches, so the walkthrough will not include code snippets; instead, it will include screenshots of LaunchDarkly.

Technical requirements

Within this chapter, we will explore how to use LaunchDarkly to work with **switches**. To follow along with the walkthroughs provided, you can use the example application from *Chapter 3, Basics of LaunchDarkly and Feature Management*.

You can find the code files for *Chapter 3, Basics of LaunchDarkly and Feature Management*, here: `https://github.com/PacktPublishing/Feature-Management-with-LaunchDarkly/tree/main/Chapter%203`. There is both a blank web application template to follow along with and a completed version of the application. To follow along with the examples provided within this application, you can use the completed version.

Exploring switches

The basic idea of a feature flag is simple and as we saw in *Chapter 3, Basics of LaunchDarkly and Feature Management*, they are easy to implement too. This simple yet powerful encapsulation has been the basis of both **rollouts** and **experiments** and ultimately, it has been about being able to enable or disable a feature before exposing it to all customers. This might have been to assess that either a **technical hypothesis** or **business hypothesis** is correct. By enabling a feature on the production environment and getting customers to interact with the new implementation, there is the opportunity to gain insight and validate an idea. Once proved successful, the feature can be rolled out to all customers. The feature flag encapsulation is then removed so that the new code is the only code that will be executed.

However, with the ability to turn on and off components within the system, there are scenarios where this ability to manage features becomes valuable permanently and not just for validating hypotheses. With feature management, it is possible to design feature flag encapsulation around pieces of functionality to be enabled or disabled as necessary. This is not a new idea, but it does become far easier to work with and manage when using LaunchDarkly than with other solutions, such as config files that might require applications to be restarted for the functionality to be toggled.

LaunchDarkly's real-time system for changing what a feature flag is serving allows rapid responses to be provided to changing scenarios in your production environment. This can help you protect the system during spikes in traffic or instability and can help with understanding specific customer journeys. In the next section, we will look at the scenarios in which **switches** can be valuable for both enabling and disabling functionality.

Switches are the first example of feature management that we will look at, where feature flag encapsulation is designed to be a long-term part of our application. While working with permanent feature flags, it is important to plan them well since, depending on what is being encapsulated, it is possible to accidentally create nested flags or dependencies across flags. In such scenarios, it can become difficult to manage the production environment. Several interconnected **switches** could result in introducing new problems to the applications. Thus, testing **switches** in both their enabled and disabled states is advised as there needs to be confidence that the **switch** is offering the planned purpose, should the need arise where functionality needs to change.

We will look at what LaunchDarkly provides to help manage permanent feature flags in the *Learning how to implement switches within LaunchDarkly* section of this chapter. Before we get to that section, though, we will look at the scenarios where **switches** can be used within your application and systems.

Discovering use cases for switches

The first type of use case I want to explore is where functionality is turned off. This approach is sometimes called the **kill-switch**, **circuit breaker pattern**, or (to give a more positive spin) **safety valve**. The only place where you might want to employ this use of switches is where you have non-critical functionality that you might want to turn off.

It is not often that there is some functionality within your system where it would be acceptable for it to not be executed. However, this approach works when the product is in a bad place and there is a need to do whatever is necessary to restore some sort of normality to the application. With this in mind, I have experience of using **kill switches** on both frontend and backend applications, such as in the following scenarios:

- **Encapsulating third-party scripts within the client**: Often, websites and applications contain several third-party scripts that are used for tracking, telemetry, and analytics. Depending on how you work with these resources, they might be controlled by people outside your team or organization, so encapsulating them within a **kill switch** can be an effective way to mitigate any risks that they pose. This is especially true in scenarios where frontend container systems are used, such as Google Tag Manager, in which a large number of tags can be added to your product via one system.

Often, these resources are important to the business but they are not mission-critical to the functionality provided by the product, so for short periods it is OK, but not ideal, for them to be turned off.

There are many ways to implement the kill switch here as each resource could have its own feature flag encapsulation around it to allow very granular control over which scripts are loaded. Alternatively, scripts could be grouped within flags for a simpler set of controls, around which resources can be turned off within the client.

This approach can be extended not just for where third-party scripts are being used within the client, but to wherever third-party dependencies are used to add value but are not essential. There is always a risk when you're reliant on third parties at runtime and by employing the **kill switch** pattern in this way, the risk of any unexpected changes on your dependencies can be mitigated.

- **Encapsulating non-essential functionality**: Another good kill switch scenario is when there is increased load or instability within a system and some non-essential functionality can be turned off. This approach to using switches to reduce the impact the load will have on your product is sometimes called *load-shedding*. Again, this is only really a viable option for non-essential functionality that is not part of the core features of the product or key business processes. There are some scenarios where even essential functionality could be encapsulated in this way, if there is an appetite to manage the risk of production systems, but this is usually a rare option.

In my experience, there might be some functionality that is demanding of the system resources that, while valuable, is not essential for the customers to be able to make use of the product. This type of feature, perhaps some form of real-time data processing that can be replayed later, could be encapsulated in a permanent feature flag to offer the opportunity for a safety valve.

While we aim to always have reliable systems, there are some scenarios where products might be dependent on unreliable and/or legacy systems. Kill switches can be effective against allowing those unreliable systems from impacting the customer experience of your product. This method works best when there are alternative ways to achieve the important functionality that those legacy systems within the kill switch encapsulation offer.

With this approach, it can be difficult to identify which functionality is non-essential as we usually only spend time and resources building essential functionality. There are usually some elements that can be turned off when dealing with a critical issue on the product. Being able to identify these components, and also ensuring there is visibility of when these features are being turned off, is important for dealing with such incidents effectively.

It is worth pointing out that you shouldn't forget to turn the features back on once normality has been restored; you might want a process around working with kill switches in this manner.

I am sure there are other use cases for **kill switches** too, but in all instances where you might be able to use this approach, you need to ensure that the encapsulation is correct. You do not want to have the switch be too broad so that too much functionality gets turned off with a single **switch** as you might never get the authorization to disable it. Therefore, testing both the enabled and disabled state of a feature is important to assure that when needed, the **kill switch** will do exactly what is expected.

It is also worth mentioning that the more granular your kill switches are, the more feature flags will need to be set up in LaunchDarkly. This is not necessarily an issue, but it does result in more time and careful management of the flags to ensure clarity on what the impact will be when toggled off. LaunchDarkly does offer some features to help manage permanent feature flags, as we will see in the next section, *Learning how to implement switches within LaunchDarkly*.

The other type of scenario where we can use the switch pattern is when we want to enable some functionality, often for a select group of customers. This approach is the opposite of the **kill switch**. This is often a less common use case as you will usually want your product to offer all the available features to all your customers. One way in which this approach can be valuable is by encapsulating verbose logging within the application, and then using a feature flag to enable the additional telemetry for specific customers. This can be highly effective when you're working to understand a customer's complaint or issue as there is far more information available about that specific user that can be analyzed. Additionally, by only collecting verbose logs for select individuals rather than all customers, the impact of collecting so much data can be managed better.

In some existing practices, it is possible to enable verbose logs, but this is across the whole application and for all customers, which can harm the performance. This extra data can result in additional bandwidth, storage, and processing being required to effectively gather the logs. Being able to selectively capture more detailed logs for select customers can remove this problem.

One interesting example of this that I have seen is to use a switch to deliver an *un-minified* version of the JavaScript files to specific users. Usually, we want to ship the *minified* version of the files to keep the files as small as possible, but this results in additional telemetry that's been gathered being removed and the file not being human-readable. By the term *minified*, I mean the condensed version of a file that is no longer human-readable, whereas the *un-minified* version is one that a developer can read and understand.

By returning the full, un-minified JavaScript file, we can gather more information and have the full names of functions to better understand where an issue might be when debugging the code. The human-readable names of functions and values can now appear in logs to speed up the debugging process.

This approach is certainly simpler to manage than kill switches as it results in very few permanent flags being added to an application's code base. This is especially if they are only used for debugging as you might only need one of these per application. With that in mind, you might find limited value in using a permanent feature flag, especially if your logs are already verbose.

Separating permanent flags from your other flags

In both cases for **switches**, how the feature flags need to be considered is different from the temporary ones that are used for rollouts and experiments. By considering these flags as a permanent part of the application, a large numbers of flags may be required within a LaunchDarkly project. This can make things seem cluttered within the project, and could result in some of the flags for experiments and rollouts remaining within the code base and LaunchDarkly for longer than necessary.

Thus, one way to manage the switches could be by using a separate project and making use of two LaunchDarkly clients within your application. This could even allow different access and permissions to ensure good governance is achieved around who should and should not be able to toggle these **switches**.

In the next section, we will look at how to use switches within your application. We will also discover how LaunchDarkly can help teams manage multiple feature flags and permanent ones within projects.

Learning to implement switches within LaunchDarkly

To implement a switch within your application, you only need to encapsulate a feature within a Boolean feature flag, as we saw in *Chapter 3, Basics of LaunchDarkly and Feature Management*. There is nothing new when it comes to the actual implementation of making use of an `if` statement to offer two pieces of functionality based on what value LaunchDarkly is configured to serve; that is, either `true` or `false`. However, it is designed to always be within your application. You may not need to make any changes to how you would implement a flag, but there might be additional documentation or abstractions within your code to provide greater context about the long-lived flag within the application.

It is worth considering how you want to work with the default value of the feature flag within your application. So far, we have only ever considered the default value to be `false` if LaunchDarkly can't be unreached by your application. However, now that we are looking to nearly always offer a feature, you might consider changing the default value to `true`. The downside that we need to understand here is that in this scenario, the feature cannot be turned off, should LaunchDarkly have any service issues. If this occurred when your product was experiencing issues, then it might be harder to get your application back to a workable state.

Once you have added the switch flag to your application, either as a **kill switch** or one that enables functionality, you will want to ensure that the switch has been configured correctly with LaunchDarkly. As a kill switch should be enabled in all scenarios apart from those exceedingly rare occasions where turning off the feature is necessary, you will want to ensure that you have all the default values set to serve `true`. The following screenshot shows how a **Kill Switch** feature flag should be configured:

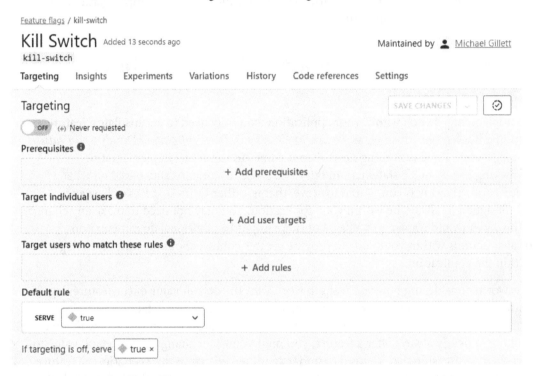

Figure 6.1 – A kill switch configured to be enabled by default

Initially, with a **kill switch**, you will want to have it configured to serve **true** for when targeting is enabled and disabled to ensure the feature is always on. When you want to turn this feature off, you can do so by toggling the value that's returned for when targeting is turned off, as shown in the following screenshot:

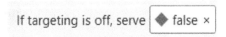

Figure 6.2 – Turning off a feature is easy with no targeting

However, as we discussed earlier in this section, you are likely to want to be able to evaluate that, in both the enabled and disabled state, the kill switch does what it is intended to do so that you can rely on it when needed. To do this, you want to have targeting enabled so that specific test accounts can experience the feature in both scenarios and validate the functionality. The following screenshot shows how to use targeting with two specific users experiencing both **true** and **false** values:

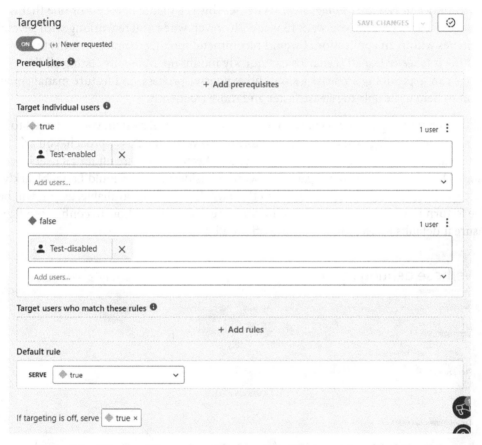

Figure 6.3 – Using targeting and specific users to test the two states of the kill switch

With a configuration like this, you can be certain that your test accounts will consistently experience the expected value from this flag to have the test pass correctly, regardless of the status of this actual flag. It is also important to keep serving true for when targeting is on or off in case someone accidentally changes the setup of this flag.

You can emulate the specific users you want to test with by setting the cookie value that we have used to consistently work with the same user. You can use the **Test-enabled** or **Test-disabled** value to assess the true or false value. The preceding screenshot shows the configuration that will cause `true` and `false` to be served for these two users.

By changing the cookie value in this way, the user will be tracked by this cookie value rather than by the GUID that we generated in the examples within *Chapter 3, Basics of LaunchDarkly and Feature Management*. Whether this is a manual process or one that is automated is based on how you want to work. However, when you're working with several kill switches within an application, I would recommend looking to automate this testing as the time it takes to test all scenarios can quickly mount up. Manually testing all the scenarios can make doing a release a slow and expensive process, and feature management is meant to help us be able to release faster and more frequently.

To set up a feature flag for the scenario opposite to that of a **kill switch**, you will need to implement a feature flag, just as we did previously. However, for this approach, you are going to want to keep the default result to `false` to keep this feature off for all users, even if LaunchDarkly is unreachable. Again, it is worth considering that should LaunchDarkly be unreachable, then the functionality from this switch will not be available, as you will be unable to turn it on for any users. The following screenshot shows how to configure a flag to ensure it defaults to keeping the feature off for all users:

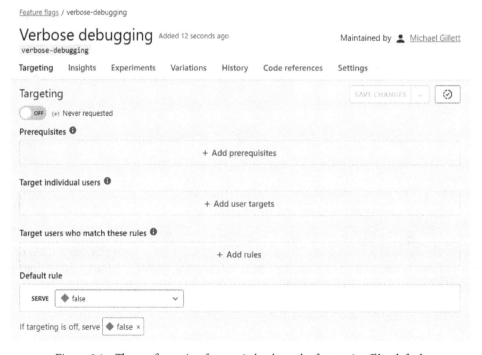

Figure 6.4 – The configuration for a switch where the feature is off by default

You will want `false` to be returned, regardless of whether targeting is enabled or not for this flag. Targeting can be used for testing, but in this scenario, it will also be needed when you're looking to turn this on for specific customers. The following screenshot shows what this switch might look like when configured for both testing and where verbose logging (the example from the *Discovering use cases for switches* section of this chapter) is needed for a specific customer:

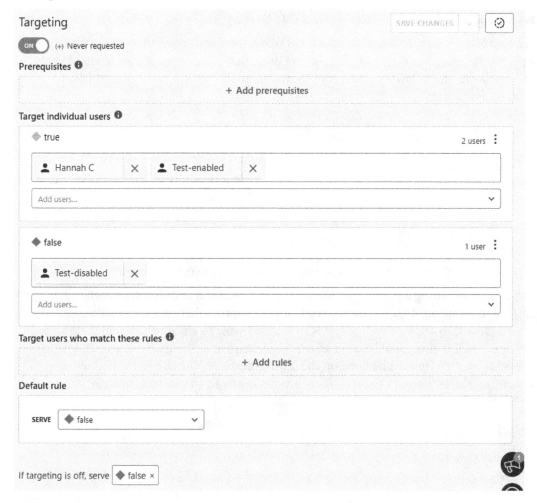

Figure 6.5 – How a switch can be turned on for just one customer

Here, a user by the name of **Hannah C** will have the verbose logging feature enabled, while the other customers will not have this feature. In the example provided previously about this type of use case, this would enable a team to view more information about the user's experience and perhaps understand the cause of a problem for just this individual.

You should now be able to set up switches within your application to dynamically disable or enable functionality to help improve stability and investigate issues. However, due to the long-lived nature of the permanent feature toggles, these can become difficult to manage. So, next, we will look at some of the ways LaunchDarkly can assist in checking on these types of flags.

Managing permanent feature flags

When creating a feature flag within LaunchDarkly, there is the option to create the flag as a permanent one. This does not alter the flag itself, but it does change how LaunchDarkly will display it. The following screenshot shows where this option can be found when creating a new flag:

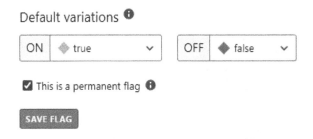

Figure 6.6 – Creating a new flag as a permanent one in LaunchDarkly

The option for setting a permanent feature flag can be found at the bottom of the configuration menu. **This is a permanent flag** is an option that goes across all environments, so once set on a flag, it is treated the same in any environment that the flag is viewed in. It is also possible to set an existing feature to a permanent one once a flag has been created. This can be done on the settings screen of the flag itself. The following screenshot shows where in the settings of an existing flag you can set it to be treated as a permanent flag:

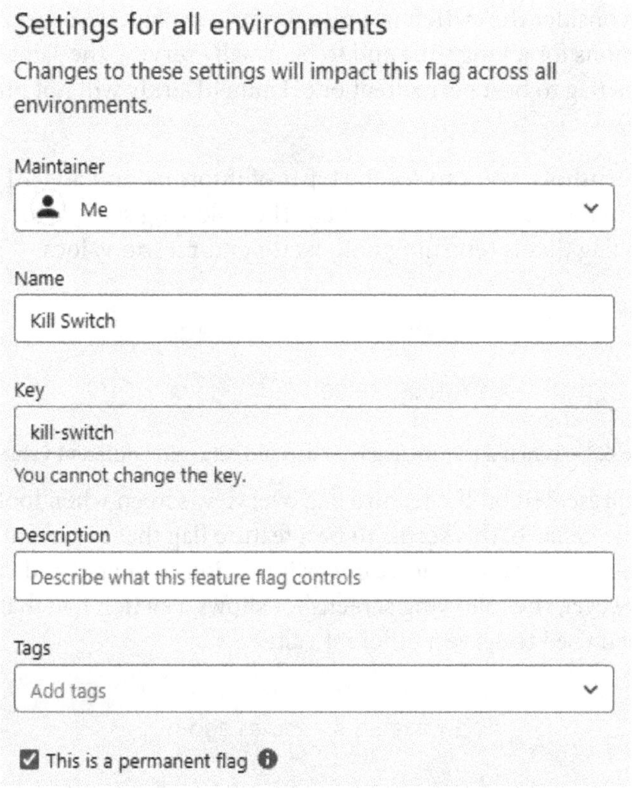

Settings for all environments

Changes to these settings will impact this flag across all environments.

Maintainer

◆ Me   ˅

Name

Kill Switch

Key

kill-switch

You cannot change the key.

Description

Describe what this feature flag controls

Tags

Add tags   ˅

☑ This is a permanent flag ⓘ

Figure 6.7 – Setting an existing feature flag to be permanent

The permanent flag setting is useful when it comes to managing flags in LaunchDarkly. Before showing you how this is useful, it is worth considering how LaunchDarkly defaults to working with flags. Due to feature flags usually being considered temporary components of an application that help deliver new code safely to production before being removed, LaunchDarkly defaults to treating flags as temporary. To help manage flags, LaunchDarkly will present information about the use of a flag, including the variations being returned by a flag and when the flag was last evaluated.

By providing this information, LaunchDarkly makes it possible to quickly identify temporary feature flags that are regularly used and are consistently serving the same value to applications. When using a feature flag to roll out a feature, this information becomes particularly useful. For example, if a flag has consistently returned true every time it was evaluated over several weeks, then it is a flag that should be cleaned up. In this scenario, LaunchDarkly will prompt that the temporary flag should be cleaned up. In *Chapter 4, Percentage and Ring Rollouts*, we mentioned that removing the implementation of the feature flag's encapsulation is the final part of the process. This does not only mean removing the flag from the source code but also removing the flag in LaunchDarkly.

However, when we consider the **switch** use case for feature flags, we expect these flags to be in our applications for a long time and to be usually serving the same value for all requests. By setting a flag to be a permanent one, LaunchDarkly will not prompt for the flag to be removed.

In the following screenshots, you can see the types of information LaunchDarkly provides to help us understand the status of a feature flag. The following screenshot shows a recently evaluated flag that is returning both `true` and `false` values:

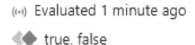

Figure 6.8 – When a feature flag is in use and returning different values

This information is presented on the feature flag overview screen when looking at a specific flag. In this scenario, this seems to be a feature flag that is currently being used for an experiment or to roll a feature out as it is being evaluated and can serve two different values. However, the following screenshot shows a **switch** flag that has been set up but has never been used to serve a different value:

Figure 6.9 – A switch that is regularly evaluated but only returns one value

This flag is being regularly evaluated but is only providing a single value to the application. As a temporary flag, this is either likely to be a brand-new flag or one that can be cleaned up. However, if it is a **switch**, then this is exactly what we would expect to see from flags in the production environment. When a flag is marked as permanent, LaunchDarkly will know that this is expected and it will not suggest that the flag needs to be cleaned up from the code base or removed from the tool.

There is also a case where a **switch** feature flag might frequently serve different values to the application. As we explored when we showed off how to configure a feature flag for a switch, it can be a good practice to run tests against both the on and off states of the switch. In this instance, the flag will have targeting turned on for specific users and will therefore be serving both **true** and **false**. Depending on how you work and with what LaunchDarkly environments the testing will be performed on, you might see differences in this flag usage data. For example, if there is a test environment that the testing of the switch functionality is run against, then you would expect to see the `true` and `false` variations being served.

On the other hand, in the production environment, it might be a test that you have decided not to run, so only the `false` value would be served, which might lead you to think you can remove this flag as it always returns the same value. In my experience, you might also want to test switches in production, but I thought I would suggest this purely as an example.

The other advantage of setting a flag to be a permanent one is that these types of flags can be filtered out from the list of flags. This helps with managing flags as you are likely to be mostly concerned with the temporary ones as features are rolled out and experimented with. These scenarios might require that the flags are interacted with frequently, so having a shorter list to scroll through makes things more efficient. The following screenshot shows how to display only temporary flags:

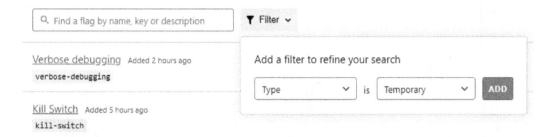

Figure 6.10 – How to show only temporary feature flags

This refined view can be saved as a dashboard for easy access the next time. At the top right of the feature flags is a **SAVE** button, which offers an overview of the configuration of the screen but also the opportunity to give it a name, as shown in here:

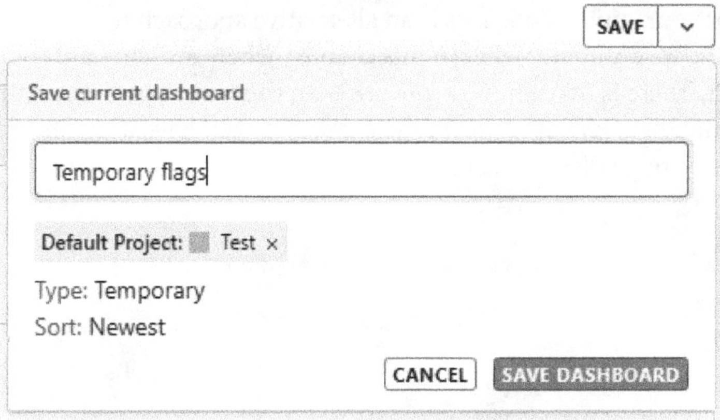

Figure 6.11 – Saving a new feature flag dashboard

LaunchDarkly will remember the dashboard you were last using and will offer that as the default the next time you access the feature flags list. The other custom or built-in dashboard can be accessed via the arrow next to the **SAVE** button. This menu even provides the ability to clone or delete any dashboard or edit the custom ones:

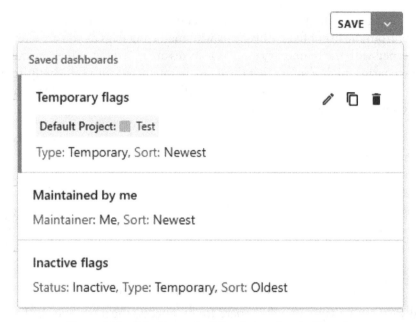

Figure 6.12 – Selecting and editing feature flag dashboards

Configuring a couple of dashboards to display a list of all **switches**, or a list that only shows your experiments and rollouts, can be an effective way to manage the permanent flags you have for your application. This is an alternative approach to working with switches to the one we outlined earlier in this chapter, where switches are confined to a separate project. There is no right way to work with them but due to the nature of switches, managing them effectively is important as they can have a significant impact on your product if they're toggled accidentally.

Summary

From this chapter, you should have gained an understanding of what a **switch** is and how feature management provides this opportunity. In many ways, a **switch** is a simpler concept of feature management than **rollouts** or **experiments**. However, a switch does not help you build and refine your product. Instead, it is used to manage your application, should there be any issues that can be improved by disabling aspects of the product and customer experience.

We explored how switches can be used to turn off functionality that is outside your control, or to deal with unlikely but possible scenarios where non-essential processes and systems can be bypassed. There might not be many opportunities to employ this strategy within your product, but this can be valuable when you want to buy some time and resources to restore normal stability and essential functionality to your customers. The other way in which switches can be used is to enable functionality for specific users, perhaps to gain insight from them.

LaunchDarkly provides the ability to work with both types of switches due to it technically being a simple use of the encapsulation of code that comes with the implementation of a feature flag. While there might not be any functionality designed specifically for a **switch** flag, there is functionality within LaunchDarkly for permanent feature flags, which we can use when we want to use switches. This allows switches to be viewed separately from the bulk of the feature flags that are used for rollouts and experiments. LaunchDarkly even treats the permanent flags differently and will not prompt to have them tidied up from your application or the tool itself.

Switches are an effective use of feature management that can prove valuable during incidents with your products. They rely on the very basics of how feature flags work and require good planning and management to achieve the best results. When used well, they empower teams and businesses to respond quickly to changeable scenarios in a way that is just not possible without feature management.

In the next chapter, we will look at **trunk-based development**. This is an approach to working with code, builds, and releases that is possible when you are comfortable with feature management. This can be an extremely fast way for you to develop new code and have it deployed to the production environment.

7
Trunk-Based Development

So far, we have explored how feature management can empower teams to separate deploying code from releasing a feature and the opportunities this presents to rolling out features, running experiments, and disabling or enabling functionality at critical moments. These are all practical ways in which feature management can impact the product, but this chapter, on **trunk-based development**, is the first time we will explore how feature management offers new practices and processes for how software development teams can approach building software.

This chapter focuses on how the practice of software development can be fundamentally changed and improved when effective management is implemented within an application. This is done through a practice called trunk-based development.

Without going into too much detail now (details will be provided in the *Understanding trunk-based development* section), the idea of trunk-based development relates to the processes around how source code/version control is achieved. This chapter assumes that you understand what **Source Code Management** (**SCM**) is and have knowledge of Git **SCM**, as I will be using this as the basis for the explanations and scenarios.

In this chapter, we will cover the following topics:

- Understanding trunk-based development

- Managing code changes using the trunk branch

- Learning how to use fewer environments

From this chapter, you will gain an understanding of what is meant by the term trunk-based development and why it can be such an effective practice for software engineers. You will explore how this can be achieved by focusing your development efforts on the trunk branch rather than working with many branches. Finally, you will learn how to use fewer environments and understand why this can speed up the deployment and delivery of new software and features.

This chapter doesn't contain code samples, nor does it have many walkthroughs within LaunchDarkly, as this is a practice that can be achieved without LaunchDarkly. This chapter will cover the theoretical ways in which Git can be used by a team that already uses feature management to follow a trunk-based approach to software development.

Understanding trunk-based development

A very common approach to building software is to use the **Git Flow** (GitKraken's definition: `https://www.gitkraken.com/learn/git/git-flow`) branching strategy to manage the source code and ensure there is good version control of the software changes. The name comes from the opportunity to make use of branches within the Git SCM tool and ensuring that changes flow through a series of branches before reaching production.

The approach was first introduced in 2010 to simplify release management by isolating different types of work into different branches. I will outline how this approach works before explaining why a trunk-based approach can be a better way of working with Git and branches. Git Flow follows the principle of having the following types of branches:

- **Main**: The main branch (often referred to as the master branch, but this is now an outdated term) is the trunk of the Git branches. This is the production-ready branch where tested and signed off code can be merged before being released to production. This is the branch we are most concerned about in the practice of trunk-based development.

- **Develop**: The develop branch is a long-lived branch that acts as a pre-production branch for other branches to be merged for testing purposes. This branch needs to be kept up to date with features, releases, and hotfix branches as it makes its way into the main branch to ensure that this pre-production version is like production.

- **Feature**: A feature branch is designed to be short-lived and is intended to allow a new feature to be built in isolation from any other changes. This is the most frequently created type of branch as this is where the bulk of new work is done. The feature branch is first branched off from the develop branch. Once code has been committed to a feature branch and the implementation is complete, the branch is then merged back to the develop branch.

- **Release**: A release branch is to be created when a new release of the main branch is being prepared. This branch is created from the develop branch and might require some minor commits. This is to ensure all the new feature branches continue to work as expected once they are all brought together, before they are merged to the main branch and released to production. Release branches need to be merged into both the main and develop branches to ensure they remain in sync.

- **Hotfix**: The hotfix branch should not be used often but allows rapid changes to be made to the main branch, should there be a need. A hotfix branch comes from the main branch and needs to be merged into both the develop and main branches to ensure that the next time the develop branch is merged into the master, the hotfix persists within the code base.

Git Flow has proved to be a highly effective way to work with multiple Git branches, but there are several problems with it, not least being that there are many branches to manage. What we have seen with feature management is that with deployments and releases being treated as separate entities, teams can move faster and release new code more often. Git Flow and its processes can get in the way of this and inhibit teams from working quickly.

Next, I will explain how trunk-based development works and how teams can work more efficiently with fewer branches. Importantly, though, this new way of working does not compromise the quality of code that is executed for the customers.

Trunk-based development

One of the key ideas behind trunk-based development is that the main branch – the trunk – should be exactly what is in production. The time between a merge into the main branch and that code change getting to the production environment should be as short as possible. With this in mind, moving a small code change through a feature branch, and then to the development branch before it gets to the main branch, can introduce unnecessary delays.

This lag between code being committed before it can get to the main branch can be made worse if there are several concurrent changes that result in merge conflicts on the develop branch, and then on the main branch. This can happen if changes are needed for features before the merge to main is complete.

With the main branch being like production, it makes sense to keep all the branches and code changes as close to this version as possible. The quicker a feature branch can be made from the main branch and merged back in it, the better it is as it reduces the chances of merge conflicts. This is because there is less opportunity for other branches to have been merged in. It also means that both the feature and hotfix branches can be considered the same thing, which, in reality, they are. In Git Flow, the reason the hotfix and feature branches are different concepts comes down to the speed at which the change needs to be made. While hotfixes need to be immediate, features can be slower. We always want to be working faster, so we want to treat all the changes with the same speed of delivery. With this in mind, the approach of hotfixes being close to the main branch is something we should be aiming to achieve for all code changes.

Trunk-based development removes the need for multiple branches to be kept in sync, in addition to everything I have already outlined, which saves time. Pull requests need to be made only when code should be brought into the trunk, rather than into both the develop and main branches.

One problem with merging new code into main too quickly is that you never want broken code in the trunk, since that is what your production system runs off. Having the develop and release branches helps offer opportunities for testing and sign-off to happen in a traditional setup. In the next section, I will explain how feature management can make merging untested code into the main branch something safe to do.

Feature management and trunk-based development

As we have seen throughout this book, good feature management allows teams to perform testing and sign off new features and changes in their production environment. This approach offers the ability to add all new code within a feature flag encapsulation and have the existing implementation as the default code to be executed. With this design pattern, it is safe to push new code to the production environment with little testing needed since once in production, no customers will experience the new code since the feature hasn't been enabled yet.

For trunk-based development to be most effective, all the changes that need to be made to the code base should be implemented in this way. This perhaps goes over and above what we've looked at in the context of feature management, since that has mostly been concerned with new components or implementations, which could be considered slightly larger pieces of work.

What is important to consider when using trunk-based development is that there is still a requirement to test your application before it goes to production. However, the only testing that's needed is regression testing. What you are looking to validate is that the current state and features of your application have not regressed with the new changes. Even with the view that all code changes are made encapsulated within a feature flag, things can be overlooked and it can be possible to alter some parts of the existing logic. These tests could be build-time unit tests or could be tests that are run following a deployment to a dedicated test environment.

With the tests being run only to validate the current functionality of the application, time can be saved by not having to update tests for the incomplete feature that is still being worked on and manually tested in production. That is not to say that the new feature should not have unit tests written for it as it is implemented – mostly, following a **test-driven development** (**TDD**) approach to writing code is one I would recommend, even when moving to trunk-based development. Having unit tests for both the old and new implementations is important and valuable. Using TDD when writing the new implementation not only helps us write better implementations but once the old implementation can be removed from the source, there is no extra work to do when it comes to tidying up the code base and maintaining a good level of code coverage by the unit tests.

There is one caveat to requiring all code changes being made in this way, and that is when it comes to bugs. Depending on the fix for the bug, it could be a simple change with a high level of confidence that the implemented fix will work, so the change can be made to the source code without feature flag encapsulation. But where a fix is larger or the confidence isn't as high, it might be good to still encapsulate the change in a flag and then enable it once it has been deployed. You would never want to make an issue worse with the changes that were intended to resolve the problem.

If many changes within the application need to be worked on at the same time in production for the whole feature to function, then each small piece can be encapsulated and make use of the same feature flag. These small pieces of code can all be released separately, and only once they are all in production does the flag need to be enabled. This can allow us to use version application functionality, which is what the release branches within Git Flow also offer. The advantage here is that this is not a single and larger release that, if it contains issues, might take a while to identify. Through delivering small pieces of code to production and having them progressively tested by enabling the feature flag for testers, it can be easier to understand what change has introduced an issue.

There is another important aspect to trunk-based development, and that is tidying up the code base to keep on top of all the feature changes being made. I alluded to this earlier in this section regarding unit testing. It is important for feature flag encapsulation to be removed from the code base as simply and quickly as possible.

Just as when a new implementation is added to the application, we do not want any aspect of it to break the existing system, so we run regression tests. Similarly, when we remove the old implementation and feature flag, we do not want that to negatively impact the new feature. For this, new regression tests can be written once a feature has been rolled out to all customers, with the tests for the original functionality being removed. Then, it is possible to deploy the release that only contains the new code and have the updated regression tests validate that everything functions before the code reaches production.

Tidying up the application once a feature has been rolled out to all customers should be something quick. It is not good if removing a feature flag and the old implementation takes as long as building the new implementation. In theory, it is possible for only the flag to be removed and some of the regression and/or integration tests to be updated for the old functionality to be successfully removed.

In the next section, we will explore some of the practical ways in which trunk-based development can be followed and how to best manage feature flags.

Managing code changes using the trunk branch

When it comes to making use of the practice of trunk-based development, there are a few ways to do so. There are more extreme options available, along with some that should be familiar to those that follow the common Git Flow methodology.

This section covers both how to make changes using trunk-based development and some suggestions for how to tidy up the encapsulated implementations.

There are two approaches to how changes can be managed and what branching strategies will be used to achieve these changes. They are as follows:

- Only use the trunk branch.
- Use short-lived feature branches.

We will explain these two approaches in the following sections.

Using the trunk branch

When using only the trunk branch, this means that changes are made directly on the main branch, without going through release branches. This can be done with commits being made directly or via pull requests and the necessary reviews. The advantage of this approach is that merge conflicts are easy to resolve, should they occur, as the branches off main and subsequent merges back, are small and frequent.. There is also no overhead of managing multiple branches and dealing with keeping many branches in sync.

Usually, the plan is that once a piece of work is merged into the trunk, then it will be deployed as quickly as possible. This maintains the view that the trunk is the same as production. The time that's taken between a change being made on the main branch and a deployment happening should be as short as possible. In some cases, the whole build and deployment can be automated once a commit or pull request is made on the trunk.

Some may be concerned that this is a very risky strategy to adopt due to the chance that unintended changes are committed to the main branch and end up in production. But this is where unit and regression testing (and potentially other forms of testing) become so important. So long as the tests are reliable and extensive, then there is the confidence that only changes that do not break the current system will get to main and production.

An advanced approach to this way of working is to use Git hooks to run unit tests before a commit is even made by a developer. Working in this way and having the tests run before the commit raises confidence in this process. Not only do we not want a breaking change to reach production, but we don't want that change in the main branch either. We can prevent this by testing even before a commit can happen.

In more extreme scenarios, teams can work very quickly and treat the main branch like production but not have the overhead that managing multiple branches can bring. By relying on robust unit tests, there can be confidence in moving quickly in this manner as it is unlikely that breaking changes get into the main branch. However, should that happen, it's very quick to revert the last merge or roll back a version as the change that was released would have been small.

Using feature branches

The other branching strategy is to use feature branches like those from Git Flow but used in a manner more like **hotfix** branches. The idea here is that it might be valuable to make several commits to a feature branch to build out a feature, before adding that to the main branch and testing it in production. This can also help prevent unexpected changes from making their way onto the main branch as more measures can be put in place, such as not allowing any commits directly to the trunk.

The way this works is that rather than making changes directly on the main branch, there is a short-lived feature branch where the code can be changed. This branch only lives until the implementation of the feature is finished and can be merged into the main branch. These branches tend to live for a few hours or days, though sometimes, they could last a few weeks.

Commits to these feature branches can be made to ensure some form of milestone is reached with the implementation before creating a pull request to get this feature branch merged into the main. At the point of the pull request, testing and reviews can be done to ensure that the system continues to work as expected. Once this feature branch has been merged into the trunk, the component can be tested on the production environment, and work can continue to be committed on the feature branch as needed. Once the feature is complete and validated in production, the feature branch can be deleted. Until the branch is deleted, there will be a need to keep the branch in sync with the main branch for any changes that have occurred.

Those who are more familiar with Git Flow may feel more comfortable with this approach as it offers many similar aspects, such as not working directly against the main branch and having the opportunity to test and create reviews before the trunk is changed. This approach is simpler and often faster to work with than Git Flow due to the reduction in the number of branches involved, the reduced chance of merge conflicts, and fewer branches needing to be kept in sync.

With these two approaches, it is possible to rely on the trunk branch far more heavily to bring about improved processes to deliver code. This is because all the new implementations can be encapsulated and can be tested more easily than when a feature flag is not used to encapsulate new implementations. Again, this relies on a key aspect of feature management, which is that deploying new code does not necessarily mean that new functionality has been delivered to customers. Taking advantage of this to improve the developer experience can be a very valuable aspect of adopting the practice of feature management.

Tidying up feature flags

As we mentioned in the *Understanding trunk-based development* section, it is just as important to be able to tidy up feature flags easily as it is to develop a feature. While trunk-based development is mostly aimed at writing and deploying code changes, we should also consider that we can use it to remove feature flag encapsulation.

Knowing how best to tidy up the code can lead to ensuring that the code is written most effectively to begin with. Thankfully, the feature flag encapsulation and regression tests we can perform on the implementation can be very useful for removing the flag.

Once a feature is deemed to have passed all the necessary tests and validations and it has been enabled for 100% of customers, the feature flag should be removed from the code base as quickly as possible. In *Chapter 6, Switches*, we saw that keeping temporary flags for longer than necessary within the application is a bad idea.

At the point where the new implementation is now the only variation to be offered to customers in production, the production tests, including regression, integration, and others, should be updated to ensure this new functionality is working as expected. It is possible to have the tests for both the old and new implementations working before this point. As we saw in *Chapter 5, Experimentation*, it is possible to test multiple variations of a feature. This could be extended not just to experiments but also to validate two concurrent implementations as one is being rolled out.

Removing a flag with trunk-based development can be done very similarly to the implementation, either with a commit straight to the main branch that removes the old implementation and the `if` statement or with a feature branch being made from the trunk to do the same. In some scenarios, there might be a few places within the code base where the encapsulation of a single flag has been implemented. This is true of more complex features. To keep the size of the commits small, there could be a new commit for each feature flag encapsulation for the implementation. This allows small iterations to be made to the code base and possible changes to production if the commits are being immediately deployed. As with implementing changes, this approach allows us to quickly identify any issues within the work being carried out to remove the feature flag.

In terms of the processes around tidying up feature flags, it is worth considering that there could be changes to how work is managed. In a traditional way of working without feature management, a piece of work can be deemed completed once the feature is deployed to production. However, where feature flags have been used, the piece of work can only be truly finished once the flag has been removed from the code base. Therefore, processes should be changed and the **Definition of Done** (**DoD**) for a piece of work should be re-evaluated to include the removal of the flag from both the code base and LaunchDarkly.

While this might not relate exclusively to trunk-based development, it is worth discussing changing the process to cater to the removal of flags. For example, for every work item that adds a feature, there could be a corresponding item to remove the flag. Both work items should be created at the same time and should have equal priority but with a sequence to them. This can work well where work is predictable to allow the cleanup work to be carried out not too long after the implementation. The downside of this approach is that if priorities change, it can be difficult to ensure that the removal of the flag will happen close to the original work item being finished, which results in a flag living longer than necessary within the application.

To get around this issue, another option is to have a piece of work that covers both adding and removing a feature flag. Subtasks can be used to break down the work involved in implementing the feature and a subtask for its removal. This at least means that once work has been prioritized, the cleanup work will go with it. The trade-off with this approach is that a piece of work might remain open for a while.

For example, when the implementation reaches production, an experiment might run for a few weeks and until a statistically significant outcome is achieved, the work to remove the encapsulation can't be done. Therefore, the piece of work remains open for several weeks, with no actual work being carried out on it. Depending on the process that's used for work management, having an item considered open until it can be removed can present a problem. This is worse if **Work in Progress/Process (WIP)** limits are used to limit the amount of work a team is dealing with. This piece of work would use up one of the items within that limit.

There's no *right way* to manage the removal of feature flags, but it is important to identify what works best to ensure the cleanup task is given just as much importance as the implementation. Without working to tidy up the code base, the application will become increasingly difficult to manage as some feature flags might become nested or conflicted. This will result in feature management slowing down the delivery time of new features rather than speeding it up.

In the next section, we will look at how fewer environments can be used when the process of trunk-based development has been adopted to write and deploy new features.

Learning how to use fewer environments for code development

In the previous section, we explored what it means to adopt a trunk-based development style and that once that is followed to a mature standard, it is possible to make further refinements to the process. While unit testing was mentioned in the previous section, the location of where to run the regression tests wasn't well detailed. That's because it is possible to move away from dedicated testing environments with this approach.

With the ability to deploy a new implementation to production, but having the feature turned off for customers, this presents the opportunity to do away with a dedicated testing environment. Code can be committed from a developer's computer to the main branch and then deployed to production, with us safely knowing that the code changes are contained within the feature flag's encapsulation. This works best when changes can't be committed until local tests have validated that nothing has regressed, as we saw in the previous section, *Managing code changes using the trunk branch*.

Once the deployment is finished, the developer can validate that the functionality they were expecting does indeed work on the production environment. Additionally, others on the team or stakeholders can also witness the progress of a feature in the production environment.

Similar to how there is an extreme way to develop code directly against the main branch, this approach can be seen as an extreme way of deploying and validating the code changes.

There are several advantages to this approach:

- **Production is the environment that the feature ultimately needs to work on**: By deploying straight to production and using this environment to validate the functionality, sign-off time is saved. It is possible that when a feature is deployed to a test environment, it results in a false positive outcome and the feature gets signed off, but once deployed to production, it doesn't work as expected. This is likely to be due to misconfiguration between the test environment and production.

  The risk here is low as there would still be testing in production before the feature is enabled to customers, but it is an inefficient way to discover bugs. Rather, if the quality assurance can be done on as a few environments as possible – just production, in this case – then there can be increased confidence that once signed off from testing, the feature is going to work. Given that testing is likely to be performed in production anyway, it is possible to skip the test environment completely. This increases the speed at which work can be deployed as testing is only done once and when a bug is fixed, it is fixed for the environment that ultimately matters.

- **Speed of delivery is increased as there are only two environments**: There is a secondary aspect to the increase in speed of deployment, even if there are no bugs, which means that the pipeline from an engineer's computer to production is smaller. There are fewer steps along the way as the application does not need to be deployed to test, with the necessary tests carried out before the same steps are repeated in production.

- **No need to keep production-like environments**: Outside of developing new features, the team can benefit from this approach by spending less time keeping the test environments in sync with production as there is only one environment to be concerned about. Depending on the complexity of the environments needed to conduct the testing, this can be a significant reduction in work.

This also means that the chance of there being discrepancies between the testing environments and production is removed, which can improve the reliability and confidence in doing releases. There would still be a need to work to ensure that local machines are configured, similarly to production, to ensure local tests are valid and useful. If there were configuration differences between local and production, then the tests before a commit or merge could yield incorrect results.

- **Costs can be reduced without the need for extra environments**: One final benefit is that costs can be reduced by removing test environments. Whether the environments are on-premises or in the cloud, the reduction in the resources will result in cost savings. This saving is in addition to any improvement to the return on investment in getting new features to production quicker by having fewer environments.

It is worth being aware that this is an extreme way of working and is not necessarily possible for all applications, teams, or testing requirements. In many cases, there will be a need to have an application pass through a test environment, where a broad range of tests can be run before deploying to production ever happens. There is also the case where there might be a need to run performance tests for each deployment, and it would be unwise to run such tests in production following a release in case the tests degrade the whole production environment.

In the case of needing to run performance tests, depending on how sophisticated the release pipeline is, it could be possible to run the tests on a test environment and then, only once validated, finish the deployment to production. If a performance testing environment was the only test environment and it did not test functionality, there would still be some benefit to having regression tests on local computers and in production. But the downside of this scenario is that there is an environment that needs to be kept production-like.

Of course, using the more traditional setup of running a release through one or more test environments is perfectly acceptable. But when using feature management within your application, you have the opportunity to not only release features in a new way but also change the way code is deployed to production. As feature flag encapsulation reduces the risk of breaking the system when deploying code, it is possible to make full use of this encapsulation and change how code is built and delivered to production. These changes can have a significant impact on how quickly teams can build new features and even introduce some cost savings too.

We have explored a lot of the theory of how moving to a trunk-based development methodology and consolidating testing environments can help development teams. Next, we will look at how LaunchDarkly itself can help.

How can LaunchDarkly help with trunk-based development?

The chapter has mostly focused on the theory of trunk-based development because it is a way of working rather than a feature of LaunchDarkly. However, there are a couple of ways in which LaunchDarkly can be useful in the context of adding new feature flags and then tidying them up.

LaunchDarkly provides a command-line tool for discovering the references to feature flag keys and where in the code they are implemented. Within LaunchDarkly itself, information about what this tool discovered is presented, including the file where the keys for a flag exist. This information makes it easier to find out whether a flag has been removed from the code base as the tool will no longer find the key in the application. The information from the tool is unable to prove that the application will or won't work in its current state, but it helps identify whether there are any references left to the key. For example, the application might not work if some aspects of a feature flag remain but the key was removed – the tool would only report that the key no longer existed. Depending on how the feature flag key and encapsulation have been implemented, this tool might provide all the information you need to know about why the flag was removed.

The tool should be added to your build pipeline. It supports several common **Continuous Integration/Continuous Deployment (CI/CD)** providers. It can also be used within a **Command-Line Interface (CLI)** and used within any pipeline if needed. Information on the supported CI/CD providers and how to install the tool can be found on LaunchDarkly's GitHub page at `https://github.com/launchdarkly/ld-find-code-refs/`. Due to the range of CI/CD pipelines available and the different platforms that the tool can be installed on, I will not explain how to install it. It is best to follow the instructions provided with the tool itself for your particular setup.

Once the tool has been installed, it can be configured to run on every commit so that LaunchDarkly can track the changing code and know when new flags have been added, or when older ones have been removed. It is recommended to install the tool within the application's build pipeline. LaunchDarkly provides information about how this tool can be configured to run in any pipeline on their documentation website: `https://docs.launchdarkly.com/home/code/custom-config`. This information can be seen on the **Code references** screen, which can be found on the **Feature flags** page within LaunchDarkly:

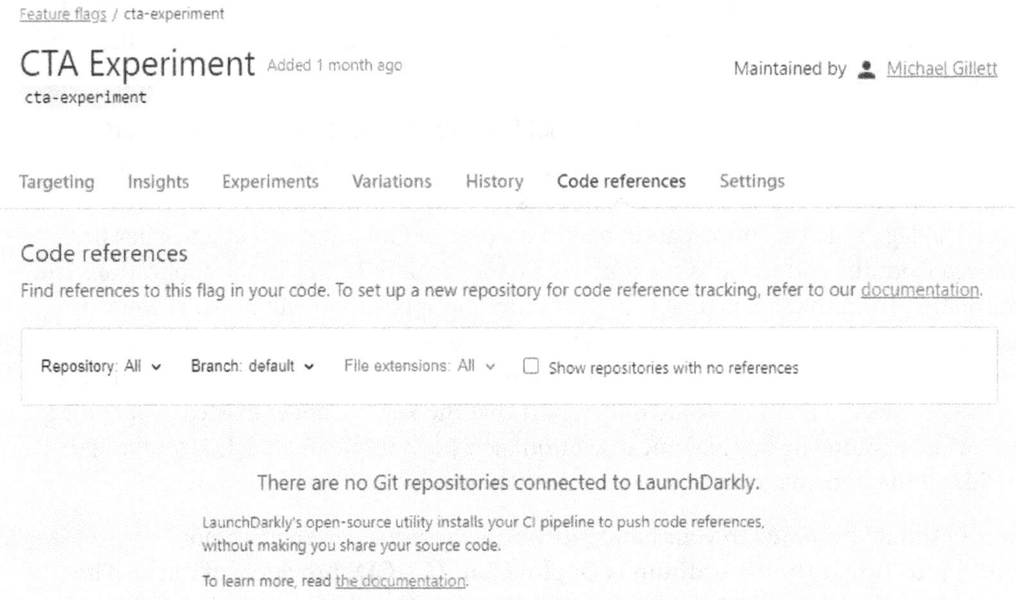

Feature flags / cta-experiment

CTA Experiment Added 1 month ago
cta-experiment

Maintained by 👤 Michael Gillett

Targeting   Insights   Experiments   Variations   History   **Code references**   Settings

Code references

Find references to this flag in your code. To set up a new repository for code reference tracking, refer to our documentation.

Repository: All ⌄   Branch: default ⌄   File extensions: All ⌄   ☐ Show repositories with no references

There are no Git repositories connected to LaunchDarkly.

LaunchDarkly's open-source utility installs your CI pipeline to push code references, without making you share your source code.

To learn more, read the documentation.

Figure 7.1 – The Code references screen of a feature flag

Before the tool is set up, LaunchDarkly provides information on how to enable it to check a Git repository.

Once the tool has been configured and run, LaunchDarkly shows the files and lines where the feature flag can be found within the code base. The following screenshot shows what the code reference looks like for the very first feature flag that we set up in *Chapter 3, Basics of LaunchDarkly and Feature Management*:

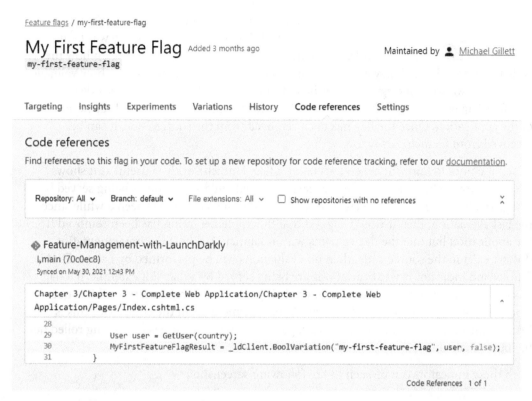

Figure 7.2 – How code references appear once the tool has found a feature flag

The screen now shows information relating to the Git repository, the commit, and the file where the feature flag reference can be found for the first feature flag. This is especially useful when you want to know where a flag encapsulation was added within the application. This is effective if a feature requires changes and the flag encapsulation exists across several files and locations and, as such, requires multiple if statements. This experience within LaunchDarkly can assist in ensuring that all the locations where the encapsulation was implemented are being tracked.

Once all the code references have been removed and the tool is run again, the **Feature flags** screen will report that the flag has been removed from the code base. LaunchDarkly itself tracks the code references, so it knows when a flag used to have one, and then the time of the commit when the flag was removed.

To reiterate, this tool does not know that the application will work, nor that the feature flag encapsulation has been completely removed. It knows where a flag key was being used within the application and some of the surrounding code. Depending on how the code is written, the tool may or may not be useful. It is worth using it to see how valuable it is to your way of working, as well as how it can help to tidy up the source code once a feature flag has been rolled out to all the customers and becomes the default experience of the application. Once the flag has been removed from the source code, it can be removed from LaunchDarkly too.

When it comes to knowing when to remove a flag, LaunchDarkly is useful as it shows two indicators: when a feature flag was last evaluated and the outcomes being served by a flag. The first indicator is the last time a flag was evaluated. If it has been a while since the last evaluation, then it would suggest that the implementation has been removed from the application but that the flag remains within LaunchDarkly. If the implementation doesn't exist in the source code, then no evaluations will be performed by LaunchDarkly. The second indicator is what variations are being served by a flag. For example, if a flag has two variations but only one outcome is being returned to customers (and it is not a permanent flag), then it would be a good one to remove from the code base. This is because only serving one variant would suggest that a new feature is now being rolled out to 100% of customers.

Both these indicators can be seen in the following screenshot:

Figure 7.3 – A flag that is only returning one outcome and has not been evaluated for a while

The blue diamond next to the **false** label shows that this flag is only returning `false` to customers and it was last evaluated 27 days ago. If an existing, temporary flag is only configured to return one value, it has probably served its purpose and should be removed from the code base. If, in a real application, a flag hadn't been called for 27 days, it would likely show that the implementation has been removed from the code base.

LaunchDarkly offers a couple of features to help us understand when a flag can be removed, as well as where in the code base it is implemented. This helps us keep on top of feature flags. When working in a trunk-based development style, being able to effectively manage flags becomes especially important as the speed of development and delivery can be faster than in the Git Flow approach to working.

Summary

From this chapter, you should have gained an understanding of how feature flags can change the processes by which software can be built and deployed. The opportunity of moving to a trunk-based development style over more traditional methodologies such as Git Flow presents some real opportunities to decrease the development and delivery time of work. There is also the chance to reduce admin overheads of managing multiple branches and multiple test environments. Once you're confident with feature management within your application, it is worth considering how trunk-based development can aid in the development and delivery of code too.

By looking at the Git Flow branching strategy, we compared the trunk-based approach against it. By relying on feature flags and keeping branches closer to the trunk rather than creating release branches and feature branches, work can be done more efficiently and without the increased likelihood of merge conflicts.

The two ways to achieve this are through making commits directly onto the main branch or by making short-lived feature branches with only a few commits, and then creating a pull request to merge it back into the main branch. While the former might seem a very risky approach to development, it is possible to move regression testing to the development computer so that bad code doesn't end up on the main branch and then in production.

On the deployment side of using feature management is the opportunity to do away with test environments. This continues the view of delivering work as quickly as possible and relying on the advantages that feature flagging presents when deploying to production. It is possible to release code directly from a local machine to production, so long as all the work is contained within the feature flag encapsulation. The new code, once deployed to production, won't be enabled for customers but can be tested directly on the environment it is ultimately going to live on.

The final part of this chapter focused on how LaunchDarkly can assist with implementing and removing feature flags from the code base. By using the code references functionality within LaunchDarkly, extra confidence can be gained that all references to a flag have been removed from an application. This functionality is complemented by some suggestions on how to manage the removal of flags from a work item and process point of view.

Trunk-based development is an added benefit of using feature management and when done well, it can enhance the delivery of a development team. This builds on the foundations of feature management and is something that can be trialed once teams are confident with feature management within their application.

In the next chapter, we will look at how migrations and testing your infrastructure can be achieved using feature management and LaunchDarkly. This will look more at backend systems and components. Up until this point, several of the examples in this book have focused more on the UI of the products being built, but feature management can be used throughout the whole technology stack.

Further reading

To learn more about what was covered in this chapter, take a look at the following references:

- Git Flow, GitKraken. This provides an overview of the Git Flow branching strategy for Git: `https://www.gitkraken.com/learn/git/git-flow`.

- LaunchDarkly "find code references" page, LaunchDarkly's GitHub. This page details how to install and use the LaunchDarkly code references tool: `https://github.com/launchdarkly/ld-find-code-refs/`.

- Custom configuration of the code references tool using the CLI, by LaunchDarkly: `https://docs.launchdarkly.com/home/code/custom-config`.

Section 3: Mastering LaunchDarkly

This last section explores in detail the full set of functionality that LaunchDarkly provides to equip you to take full advantage of feature management and LaunchDarkly. From managing and configuring feature flags, users, and experiments through to understanding how the debugger works, there is a wealth of information and functionality that LaunchDarkly provides and this section goes through all aspects in depth. To conclude, this section details the features available for managing the access, authorization, and security of your LaunchDarkly account to allow you to confidently use LaunchDarkly to make changes to your production systems.

This section comprises the following chapters:

- *Chapter 9, Feature Management in Depth*
- *Chapter 10, Users and Segments*
- *Chapter 11, Experiments*
- *Chapter 12, Debugger and Audit Log*
- *Chapter 13, Configuration, Settings, and Miscellaneous.*

8

Migrations and Testing Your Infrastructure

So far, we have mostly been concerned with how feature management and LaunchDarkly can be used to deliver new functionality to the frontend experiences through **rollouts** and **experiments**. In this chapter, we will look at a different way in which feature management can be used on backend systems. When used in this way, feature flags can be used to help migrate entire systems and even offer the opportunity for new types of testing.

We will build on what we learned in *Chapter 4, Percentage and Ring Rollouts*, to understand how a **migration** can be performed. In that chapter, I touched upon some of the scenarios that will be looked at in greater detail within this chapter.

Many of the examples provided so far have helped you make relatively small changes to applications, especially when it comes to running experiments, but there are occasions where much larger architectural changes need to be made to systems. Feature management can also be used to ensure that controlled and safe migration occurs when migrating from an existing design to the new infrastructure. Within this chapter, we will look at how this can be achieved and some of the lessons learned.

When considering using feature flags within your infrastructure, some new testing opportunities present themselves. This is, again, within the scope of testing in production but offers insight into the behavior of the systems rather than how the customers interact with your product.

In this chapter, we will cover the following topics:

- Discovering how to perform migrations with LaunchDarkly
- Learning new ways to test your new infrastructure

By the end of this chapter, you will have gained an understanding of how feature management can be used on backend systems and understand that it is not just a methodology for releasing and experimenting with frontend features. Through using feature flags in this way, substantial changes to entire systems and processes can be made safely, and testing can be performed on the infrastructure that would otherwise not be easy to run.

> **Note**
>
> Within this chapter, there will be no code samples, but there will be some walkthroughs of how to achieve the migrations and tests within LaunchDarkly.
>
> In the descriptions within this chapter, I will use the term *client application* to refer to the system that is calling a backend system. This is not necessarily a frontend application but one that has a dependency on a backend system.

Discovering how to perform migrations with LaunchDarkly

Much of what has been described so far regarding feature management and where LaunchDarkly can be useful has been about making small, testable changes to your application, such as rolling out the change to a segment of your customer base and validating that it works technically, before moving on to running an experiment to see how the customers interact with and experience the feature. Sometimes, the changes that need to be made just cannot be done in such a small or iterative manner. Instead, there is a substantial change that needs to be released to production.

These large changes could include the following:

- Rearchitecting existing services. That requires working within client applications to support this change, which could result in fewer dependencies within a client application.

- Rebuilding a single backend system, which requires users to be migrated.

- Moving systems to a new location or URL. For example, this could be moving a service from on-premises to the cloud.

Thankfully, even with large changes, feature management and many of the aspects that we've already looked at within this book are still applicable. Some things are worth considering when planning the implementation of larger and wide-ranging changes within your system:

- Firstly, it is worth being aware that to make use of feature management for infrastructure changes and migrations, there is a need to have both the existing and new systems able to run in parallel. This is similar to having both the old and new implementations of a feature within the code base. This allows for a safe and controlled move from one system to the other when the migration needs to happen.

- Secondly, depending on the scale of the change, it could be that multiple apps need to move to the new infrastructure at the same time to ensure data concurrency, accuracy, and so on. For this scenario, each of the client applications can have the same feature flag within them, so they can all be managed with a single flag configuration change. An alternative approach to this is to use **prerequisite** targeting, which we will explore in the *Using separate flags to perform a migration* section.

- Also, when working with large migrations, it is important to have each client application treat requests from the same customer in a consistent manner. In *Chapter 3, Basics of LaunchDarkly and Feature Management*, we looked at how to consistently track a user across multiple page loads through a GUID stored in a cookie. This approach can be followed for these migrations too.

  By using the same GUID value across both the frontend application and any backend systems, it is possible to ensure that a customer is treated as the same user within LaunchDarkly. This would require that the GUID value be passed between all systems and used as the unique identifier for that customer when creating the LaunchDarkly User object within each application.

- The GUID methodology works well when dealing with anonymous users. However, if the infrastructure migration is affecting logged-in users, a different approach can be taken. For example, if a user has an account, then their username or a unique identifier could be used. In either case of dealing with known or anonymous users, when conducting large migrations, it is important that multiple applications and systems can consistently identify the customer.

We can use feature management and LaunchDarkly for system migrations by encapsulating the endpoints of the backend system that are about to change within the client application. These endpoints could be APIs, database connections, or other integrations that need to be implemented in the other system. Often, when a change is made to the endpoint, there is a requirement to send more data or support a new data structure within the response. The implementation to support these new request and response parameters needs to be encapsulated alongside the call to the endpoint itself. In this migration example, where a new architecture would consolidate the number of dependencies a client application has, would involve work to use new URLs and also rewriting the code itself to connect to fewer systems.

There is also the likelihood that some new configuration that exists outside the feature flag encapsulation would be required. For example, a database connection string that needs to be kept secure could be added to the application at build time to prevent it from being read by anyone looking at the code base. When moving from one database to another, there will be the need to have both connection strings within the application until the migration is complete.

The point of adding extra logic, endpoints, and/or configuration is to make it possible to connect to either the old or new part of the infrastructure by changing just one feature flag. The application should be able to work with the flag enabled or disabled.

Once one client application can be proven to work with both old and new systems, similar work can be done in any affected client application. Sharing the same feature flag can simplify the migration, but it possibly presents the need for flags to be shared across applications, teams, and projects.

At the point at which all client applications have encapsulated the implementation for the new infrastructure, including the new endpoint and handling any changes to the request, then it would be possible to test that everything is working as expected. Just as we saw in *Chapter 4, Percentage and Ring Rollouts*, the implementation should be tested internally to validate everything functions as expected. With the increased impact that a bug within the implementation could have on the system when the feature flag's configuration is changed, extra due diligence is advised.

When we explored the technical testing of feature flags in previous chapters, the scope was small. However, with migration, there are several additional things to consider, such as the following:

- The new or updated backend system itself
- Ensuring each client application is interfacing with the new backend system correctly and continuing to work as expected
- Ensuring that each client application continues to work together correctly

Once validated, the feature flag can be updated to roll out the new system to customers. By having all the apps treat the customer in the same manner, whether their requests should go through the old or new system, it is possible to slowly migrate from the old to the new. How a migration is rolled out depends on the actual change to the infrastructure, but it can often be possible to segment the customer base so that not everyone gets the new version at the same time. As ever when using feature management, it is best to avoid big bang releases, so being able to segment the user base in this manner helps.

I have seen a new authentication system rolled out where customers in different countries were migrated off the old version. This migration required several frontend and backend applications to encapsulate all the new authentication logic and requests, and all of this was managed with a single flag. It took a while for the entire system to be successfully validated but once completed, the actual migration of customers was relatively painless and was done in a controlled and methodical way.

However, in some cases, the migration needs to go from 0% to 100% of customers at the same time. In that case, extra testing would be advised before exposing the change to customers.

Once the migration is completely rolled out to customers, it needs to be removed from all the code bases. This process is often more involved than with a small feature flag. Firstly, there will be many applications that will need to be updated to remove the encapsulation. Secondly, depending on how the backend system connection has been done, there might be a need to remove some code or configuration that is outside the feature flag's encapsulation. For example, as we saw earlier in this chapter, there can be scenarios where there is a need for multiple database connection strings that are added during build time; this process is outside of the flag itself.

As we saw in *Chapter 7, Trunk-Based Development*, it is possible to use the **Code References** tool from LaunchDarkly to detect all the code references of a feature flag. This can be especially useful when we need to clean up a flag from multiple applications. Once the flag is no longer being evaluated, it can be removed from LaunchDarkly.

Before we look at the types of testing that can be performed when adding feature management to backend systems, I want to highlight the complexity of working with feature flags across multiple applications. It is likely that with large migrations, flags will be shared across applications and teams. When this happens, it can be easy for applications to nest or conflict with flags that they are not in complete control of, and this migration flag could end up living within a code base for some time. I have seen cases where additional logic is built within certain shared flags, which causes several issues. A central team might want to change the targeting of the shared flag, only to find that it now breaks an application. So, it is important to realize that there is value in shared flags, but that they should only be used infrequently.

With several teams all working against the same flag, some applications might be updated quickly to introduce this encapsulation, while others will take much longer. This should be considered when the work is being planned to ensure that work is delivered to production around the same time. Without this planning, some teams might deploy work to production long before it can ever be enabled to customers. This results in an inefficient process. It would have been more valuable for the team to have focused their time and effort on work that could become customer-facing sooner.

Using separate flags to perform a migration

An alternative approach to sharing flags is to allow each application or team to have its own flag, and then ensure that there is good scheduling between the teams for when the flag's configuration should be changed. In this case, it is possible to use LaunchDarkly's **prerequisite functionality** within the targeting of a flag. This allows the outcome of other flags to be used in the evaluation of another.

With this approach, the flags could be managed in a way where the frontend application evaluates whether a customer should be migrated or not. The subsequent flags, perhaps on other frontend applications or backend systems, can then use this flag's evaluation as their own. The following screenshot shows how this type of configuration can be set up in LaunchDarkly:

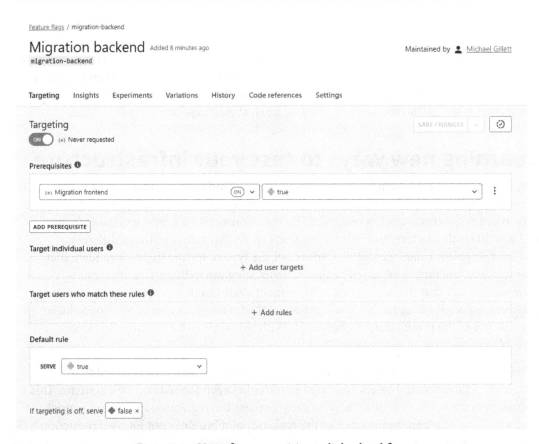

Feature flags / migration-backend

Migration backend Added 8 minutes ago

Maintained by 👤 Michael Gillett

`migration-backend`

Targeting   Insights   Experiments   Variations   History   Code references   Settings

Targeting   SAVE CHANGES ⌄   ⊘

🔘 ON   (+) Never requested

Prerequisites ❶

(+) Migration frontend   (ON) ⌄   ◆ true   ⌄   ⋮

[ADD PREREQUISITE]

Target individual users ❶

\+ Add user targets

Target users who match these rules ❶

\+ Add rules

Default rule

SERVE   ◆ true   ⌄

If targeting is off, serve ◆ false ✕

Figure 8.1 – Using flag prerequisites to link related flags

In this example, we have two migration flags: one for the frontend migration implementation and one for the backend changes. The preceding screenshot shows the backend migration flag setup with the **Prerequisites** settings for the frontend flag's **Migration frontend** value being **true**. Because the targeting configuration is on the backend flag, whatever the frontend migration flag is evaluated to will be the same on the backend system. The client app is free to turn on the migration and target users as needed as the backend system will follow suit. This approach still relies on the customer being recognized as the same person throughout all systems.

If a request makes it to this feature flag without the client application having been evaluated to **true**, then the default **false** value will be returned. This is another instance where the pattern for implementing feature flags should only execute new code when the flag is **true**. We looked at this pattern in *Chapter 3, Basics of LaunchDarkly and Feature Management*.

Migrating from one system to another can take a large amount of work, but feature management can make the process much less risky and, through testing the new implementation internally before any customers, can add extra reassurance. However, while feature flags and LaunchDarkly can be used on both frontend and backend systems, some new forms of testing can be done in addition to standard functional, regression, and integration tests. In the next section, you will learn about these other opportunities.

Learning new ways to test your infrastructure

Once feature flags have been implemented on the backend systems, in addition to the frontend applications, some new testing opportunities can make use of controlling what downstream systems are being requested. By this, I mean that a client application's use of LaunchDarkly can target a different endpoint or configuration so that a different backend service is called for testing. This is not the type of testing that will validate that components function as expected or that systems integrate well; instead, this testing relies on the fact that the production environment itself can be safely tested. In previous chapters, especially *Chapter 5, Experimentation*, the focus was on being able to evaluate the features of the production environment. With this approach to testing, it is also possible to assess the infrastructure itself.

When discussing performing migrations in the previous section, we explained that different endpoints could be encapsulated to move between the old and new systems. This approach can be used in another way too, where there is a regular endpoint and a mock endpoint. The mock endpoint acts like the real endpoint but does not hit the production backend systems.

There are many ways that the mock endpoint could be set up, as follows:

- Putting a mock response within the feature flag encapsulation so that when we want to hit the mock service, no request is made. Instead, there is a hardcoded implementation of what the request and response would be like. This implementation does not allow us to test the requests being made, which can restrict some of the types of testing that can be executed with this approach. For example, the tests would not uncover an issue in the implementation of a request being made or the response being returned; it would only show issues with how the data was being used from another endpoint.

- Creating a separate application or function endpoint that returns a hardcoded value with no execution of logic. This way of creating mocks can be more complex and possibly expensive, but it does allow us to test the request from the client application. This approach addresses the downside of putting a mock response within an application itself.

Creating mock services and responses in these ways is not a practice that is exclusive to feature management, and it is something worth considering as a best practice anyway. One advantage of combining mock tests with feature management is that it can be easy to run tests on production, without the need for environment or configuration changes to run the tests.

These tests are most commonly implemented as a temporary exercise to ensure systems will work for key events or product launches. Having these mock endpoints as permanent parts of applications could be an effective way of assessing the infrastructure on a long-term basis, but it would result in many long-lived feature flags, which can become difficult to manage, especially if they end up nested within other flags.

Once the mock endpoint has been implemented within an application, the default behavior should be to hit the real endpoint. After all, we will want customers to experience the real functionality. We can then turn on the mock endpoint logic for certain types of requests, such as those that come from the testing team or automation tests.

We can use some of the attributes we saw in *Chapter 3, Basics of LaunchDarkly and Feature Management*, plus others if we provide them within the LaunchDarkly initialization in the application, such as the following:

- IP address
- Browser
- Device operating system
- Specific user ID
- Custom query string parameter

Targeting the mock endpoint against these types of attributes allows for the tests to be correctly identified and provided with the mock implementation. For example, automation tests could always come from a certain IP address, so enabling the mock endpoint for those requests would be easy to configure. If specific people wanted to test the mock endpoint, then targeting users makes the most sense. On the other hand, if it is not very easy to categorize when the mock endpoints should be used, a system could be implemented to provide a custom query string value within the URL that is passed through to LaunchDarkly, on the User object's initialization, to enable the test.

These tests prove that the production environment can deal with several non-functional requirements that are difficult to prove otherwise. In some cases, a performance environment can be used to perform these tests, and this keeps the extra traffic away from the production environment. However, as we have seen with other use cases of feature management, such as experimentation, it can be advantageous to ensure things perform on production as expected. This is the environment that ultimately matters, so proving things work here is the most valuable place to run the tests. There is also the time- and cost-saving aspect of not needing to maintain additional environments.

Some of the tests that can be performed on production using feature flags include the following:

- **Load test**: This is where a large amount of traffic can be sent to a production application in a short space of time to ensure that the system continues to function, even with large volumes of requests. Using the mock request and response for these tests means that no load is put on to other parts of the wider system. This allows each application and component within the entire system to be tested with a high load and signed off, stating that it can handle the traffic to an expected level.

- **Soak test**: This test puts increased traffic on an application for a prolonged period. The traffic is not at the same level as a load test but will put extra strain on the resource. This test can be used to identify memory leaks and similar issues where, over time, the application's performance might degrade. Again, the mock endpoint can be enabled for tests so that this high level of traffic doesn't impact downstream resources.

There might be a concern with this approach as running these types of tests in production could introduce instability that affects your customers. In a comparable way to some of the approaches we outlined in *Chapter 7, Trunk-Based Development*, this is an extreme way of testing the production infrastructure and one that only works with good governance, and if you are confident in utilizing feature management. Identifying when to run these tests is important, as is being able to see the performance of the system in real time. Should the performance not hold up as expected, then the tests need to be stopped immediately to avoid this impacting the customers. However, when the tests are well timed and the systems hold up, there is extra confidence in how well the production environment will deal with high levels of traffic.

The types of tests we've outlined are by no means the only ones that could be run using feature flags in this way. For example, it could be that, in the mock endpoint implementation, some delays have been added within the code: now, the mock test could be used to emulate latency within the requests and the response to validate how the application would handle this. It could even be possible to run a multi-variant test, where multiple delays have been implemented. This would provide a greater understanding of where latency starts to bring instability into the experience.

This is another example where feature management can empower teams to explore testing in production. Just like testing individual features, being able to test the infrastructure that the system runs on can provide valuable information and yield efficiencies in reducing the number of environments needed.

While LaunchDarkly does not provide any specific functionality for performing these infrastructure-level tests, feature flags can be configured to enable the new backend requests as needed. *Chapter 4, Ring and Percentage Rollouts*, has already outlined how a feature can be rolled out to specific users, or groups, or by targeting specific attributes.

Summary

In this chapter, we looked at how feature management can be used in more ways than just building features. By using feature flags in a slightly different scenario, it is possible to make large system changes and migrate from an old infrastructure to a new one in a safe and controlled manner.

We looked at how, by encapsulating the new system's endpoints, updated requests and responses, and new implementations, migrations can be safely performed. This work can be done in many client applications at the same time for larger migrations. Once the implementations have been tested and validated, it is possible to migrate customers from the old systems to the new ones.

Building on the approach of adding feature flags to backend systems, there are some new opportunities for testing the production environment. Through building mock endpoints, a high level of load can be put on the product to ensure it handles traffic to the expected levels. By testing production in this non-conventional manner, extra confidence can be gained in terms of how well the product can cope with high load.

This chapter should have, once again, shown how versatile feature management can be and what new opportunities it presents. As we have seen in previous chapters, using feature flags within applications offers the ability to carry out significant changes quickly and safely, without the need to have a deployment and a change being made to the application happen at the same time.

This concludes the second section of this book, in which the core concepts of feature management and how to implement them using LaunchDarkly have been covered. The basis of feature management includes how to **roll out** features to **percentages** or groups of customers, through to running **experiments**, using **switches**, and adopting a **trunk-based development** approach to software development.

In the last section of this book, we will look at LaunchDarkly itself to enable you to get the most out of the tool, as well as ensure that you and your team can maximize the benefits of feature management. In the next chapter, we will take a detailed look at all the features and functionality LaunchDarkly provides for creating, managing, and removing feature flags. We will also look at all the data and insights the tool offers for understanding how a flag and an experiment are performing within applications.

9
Feature Flag Management in Depth

This chapter is the start of the final section of this book, in which we will explore the full LaunchDarkly product. This will empower you and your teams to take full advantage of everything LaunchDarkly offers and ensure that it is well managed and configured correctly.

This section assumes that you already have an account, project, and environment set up within LaunchDarkly. *Chapter 3, Basics of LaunchDarkly and Feature Management*, provides all the information you will need to set up LaunchDarkly.

In this chapter, we will explore all the features and functionality we can use to manage and configure feature flags within LaunchDarkly. So far, we have looked at some aspects of the available components within the tool, but not all of them. This chapter will ensure you can make the most of LaunchDarkly and even discover new ways to use the tool for your own needs that I have not outlined in the previous chapters.

In this chapter, we will cover the following topics:

- Exploring LaunchDarkly – the dashboard and feature flags
- Working with targeting and variations
- Discovering insights and history
- Learning about experimentation
- Using code references
- Exploring Settings

Throughout this chapter, there will be screenshots and walkthroughs to show you how to use the full LaunchDarkly product. Some of the functionality covered has already been covered in other chapters. References will be provided where more detailed examples or code samples have been used previously within this book.

Exploring LaunchDarkly – the dashboard and feature flags

To begin with, we will start with the feature flag dashboard, which is the default screen of LaunchDarkly. It is within this screen, besides the individual feature flags, where most of your time will be spent within LaunchDarkly.

Overview of LaunchDarkly

The following screenshot shows the whole LaunchDarkly experience. We will explain this before we provide information about the functionality available to manage the flags on the dashboard:

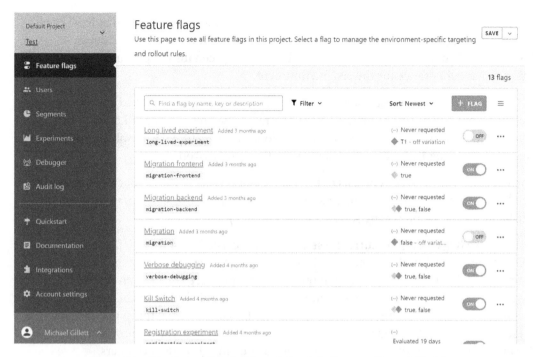

Figure 9.1 – The feature flag dashboard, the default page of LaunchDarkly

On the left-hand side of the LaunchDarkly product is the sidebar, where you can navigate between the main areas of the site. The large area on the right is where you will find the main functionalities of the tool, and this changes for each area of the site. In the preceding screenshot, you can see a list of feature flags. We will come back to them soon, but I will briefly explain the components of the sidebar.

The important thing to be aware of when working with LaunchDarkly is which project and environment you are in. Features flags are created within a specific project, so ensuring you are in the right one is important. Once created, a flag exists in all the environments that have been set up for a project, which makes knowing which environment is selected important so that changes to production do not accidentally occur. How to configure these two settings will be covered in greater detail in *Chapter 13, Configuration, Settings, and Miscellaneous.*

At the top left of LaunchDarkly, you can see the project and environment that have been selected. In the preceding screenshot, this is the large box at the top of the sidebar. If you access LaunchDarkly, by default this box will be green for the production environment and orange for the test environment.

In *Chapter 3*, *Basics of LaunchDarkly and Feature Management*, we looked at projects and environments in detail, but it is worth reiterating the importance of knowing which project and environment you are in before we look at the feature flag dashboard. For this chapter, I am using **Default Project** and am in the **Test** environment.

The LaunchDarkly sidebar consists of the main components of the product, configuration, and help resources and allows you to access your account settings. The gray panel on the left-hand side of the preceding screenshot shows the sidebar.

The **Feature flags** section is currently selected in the preceding screenshot. The other areas of the site within the sidebar will be explored in the next few chapters. Underneath the horizontal divider, there are useful resources and product information that won't be covered in this book. The resources and information provided by LaunchDarkly for these sections are updated frequently and cover a lot of the same areas of this book. Due to these changes and not being the functionality of the tool, I will not go through them. The **Account settings** and **Profile** sections will be covered in subsequent chapters.

Looking back at the preceding screenshot, most of what is shown is of the feature flag dashboard. We will cover that next.

The feature flag dashboard

Often, many of the projects you will work with will have several feature flags at any given time, and being able to manage these flags can become tricky. Fortunately, LaunchDarkly offers several features to help with this.

First, there is its search and filter system, as shown in the following screenshot:

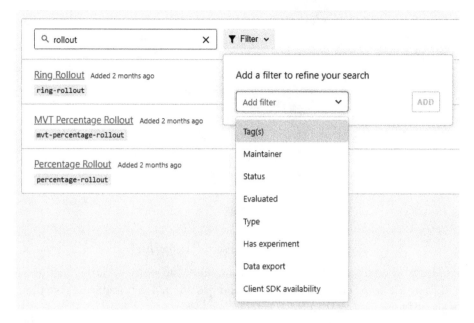

Figure 9.2 – Search and filter system for feature flags

The previous screenshot shows a search for the term **rollout**, along with the **Filter** menu and the attributes that can be used to refine the list of flags. For example, it is possible to show only those flags that have been evaluated in a given time frame, as shown in the following screenshot:

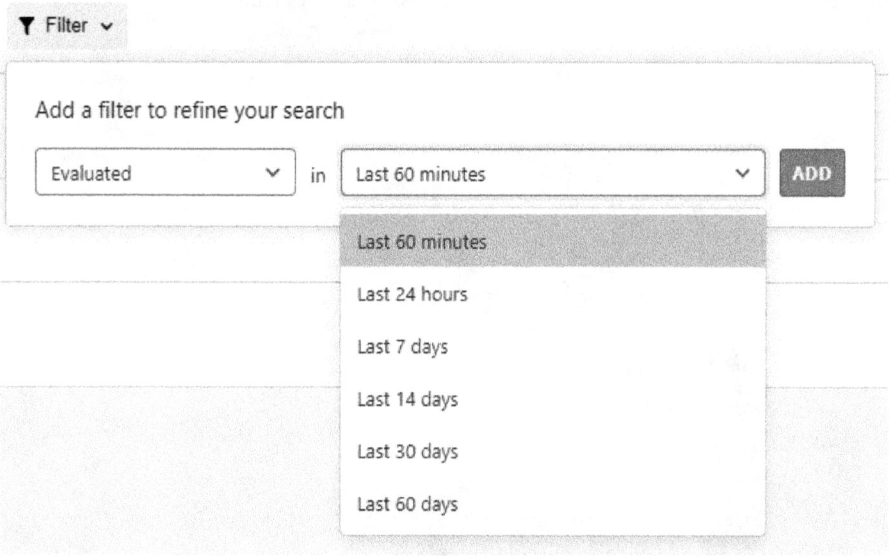

Figure 9.3 – An example of the filtering options available

Once a list has been refined, it might be that the ordering should be changed to make it easier to find the relevant flags. The following screenshot shows the **Sort** options in LaunchDarkly:

Figure 9.4 – The options for sorting feature flags

With the options to search, filter, and sort, you can tailor the view of feature flags as required. If there are some configurations that you use often, it is possible to save this view of the dashboard for easy access in the future.

Working with dashboards

By default, LaunchDarkly has two default dashboards. One is to view all the flags that you maintain, while the other is to show all the inactive flags within a project. The functionality of saving a new dashboard and accessing the available ones can be found at the top right of the screen. The following screenshot shows this menu:

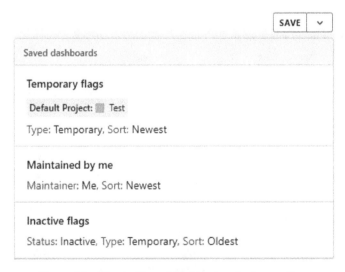

Figure 9.5 – The dashboard for viewing the feature flags

The **Temporary flags** dashboard is the one we created in *Chapter 6, Switches,* and helps when we're working with several temporary and permanent flags within a project. The dashboard is attached to the specific project and environment that was being viewed at the time. To save a new dashboard, click the **SAVE** button in the top right. This will present an overview of the configuration to be saved and the opportunity to name the dashboard, as shown in the following screenshot:

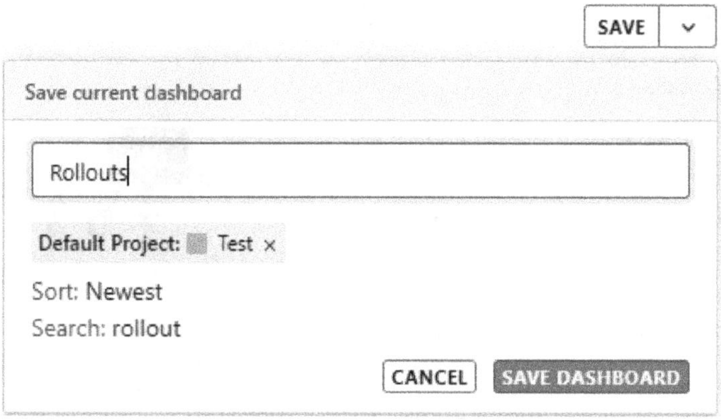

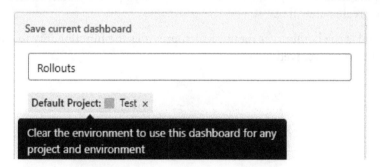

Figure 9.6 – Saving a new dashboard

By default, the dashboard is attached to the given project and environment. However, it is possible to create dashboards for all projects and environments, as shown in the following screenshot:

Figure 9.7 – Making a dashboard available for all projects and environments

By clicking the **x** button next to the project and environment, the dashboard will be saved so that it can be used across all setups. If you have created dashboards that are only for specific environments or projects, the dashboard configuration can be edited. You can edit a dashboard via the dropdown menu, as shown in the following screenshot:

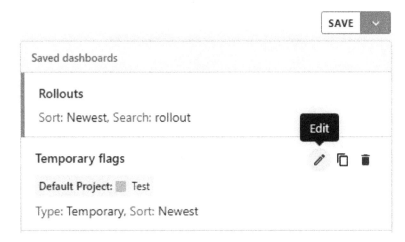

Figure 9.8 – How to edit a dashboard

Editing a dashboard allows you to change its name, as well as remove a project or environment restriction. Dashboards can also be copied or deleted. There are two remaining aspects of the feature flag dashboard I have not covered yet:

- One is quite simple, but perhaps the most important part of LaunchDarkly – the ability to create a new flag.

- The second is the hamburger menu, which provides the functionality to compare flags and to view archived flags.

Both are detailed next.

Creating a feature flag

When clicking the + **FLAG** button, you will be presented with all the options to create a flag. A quick look at this was provided in *Chapter 3, Basics of LaunchDarkly and Feature Management*, but we will now take a more in-depth look at this. The following screenshot shows the top portion of the panel, which is where we can create a new flag:

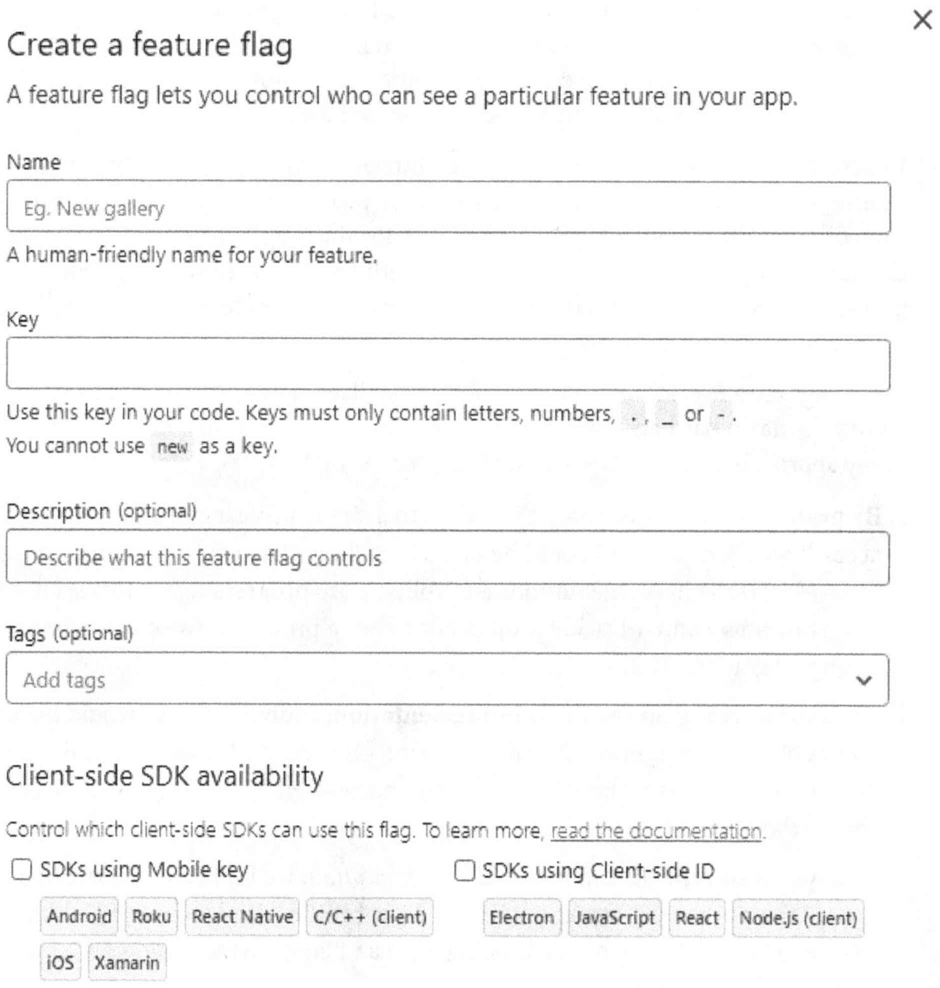

Create a feature flag   ✕

A feature flag lets you control who can see a particular feature in your app.

Name

Eg. New gallery

A human-friendly name for your feature.

Key

Use this key in your code. Keys must only contain letters, numbers, `.`, `_` or `-`.
You cannot use `new` as a key.

Description (optional)

Describe what this feature flag controls

Tags (optional)

Add tags

Client-side SDK availability

Control which client-side SDKs can use this flag. To learn more, read the documentation.

☐ SDKs using Mobile key   ☐ SDKs using Client-side ID

Android   Roku   React Native   C/C++ (client)   Electron   JavaScript   React   Node.js (client)

iOS   Xamarin

Figure 9.9 – Creating a new feature flag (the first portion of options)

Here is a list of the important fields:

- **Name**: The name of the feature flag can be anything you like; it is advisable to adopt a naming convention for your flags. Having consistent naming for the feature and purpose of the flag (rollout, experiment, and so on) can help when it comes to managing flags.

- **Key**: As we have seen in previous chapters, the key is important as this is what is used within the application to find out which flag should be evaluated. Again, though this can be anything you like, a naming convention might be useful to help you manage the flags effectively within your code base.

- **Description**: A **Description** is not needed but can be useful, especially when dealing with large teams or many flags within a project. The more information someone can read about a flag, the easier it is for that flag to be managed correctly, although that person does not work closely with the application or flag. For permanent feature flags which might live in an application for years, having a description is especially useful.

- **Tags**: Tags are a useful way to view and manage flags. They can be used to filter flags within the dashboard and then have a saved view of a certain list of them. There are many approaches to how tags can be used; for example:

  a. **By project**: This is where flags that relate to a single project can all be tagged accordingly. A dashboard could be saved that shows flags for a particular project to show how the implementations and rollouts are progressing. In this scenario, I don't mean a LaunchDarkly project but rather a project of work involving multiple features.

  b. **By those working on the flag's implementation**: Individual flags would have the names of the developers and testers working on them added as a tag. This can help if questions arise about the feature being worked upon as it would be clear about who we can talk to.

  c. **By hypothesis type**: In *Chapter 5, Experimentation*, we looked at the two types of hypotheses that can be experimented with: business and technical. Adding a tag stating which type of test is being run can help provide insight about the purpose of the feature flag.

There are many other ways you can use **Tags**, and changes may be required once a flag has been created.

The **Client-side SDK availability** section allows us to have flags working via the client-side or mobile SDKs. By default, flags work with the server-side SDK. The reason for this is that there are security concerns and cost implications of running LaunchDarkly flag evaluations in a client application.

The security concern is that the SDK key needs to be available within the client app, which increases the likelihood that it leaks. If that were to happen, then a bad actor could view all the flag evaluations for any customer. By ensuring that flags are enabled for these client SDKs, it adds protection against them being accidentally used without adequate measures being taken. In my experience, it is best to use the server-side implementation. There might be cases where this is not possible, but the choice to use client-side SDKs should not be taken lightly.

Next, we will look at the remaining options for creating a flag, as shown here:

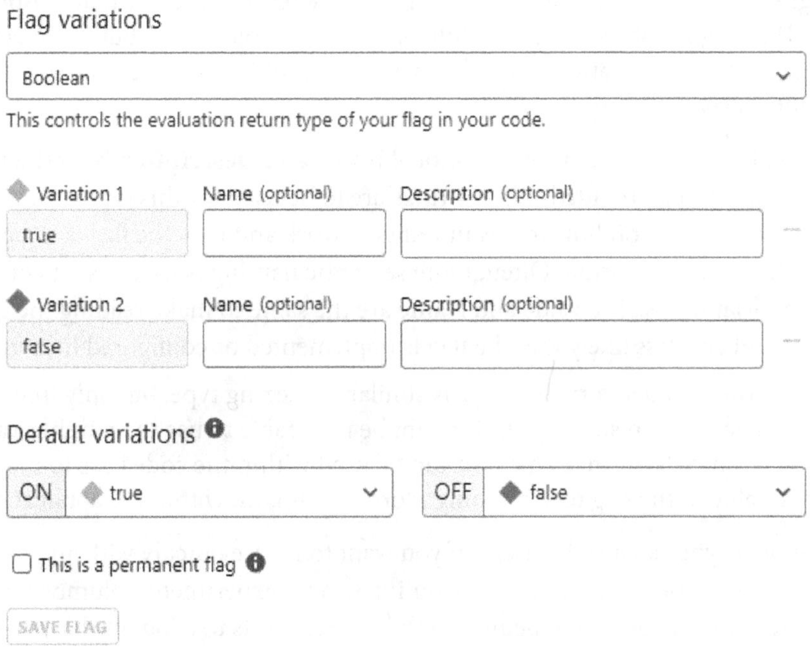

Flag variations

Boolean	⌄

This controls the evaluation return type of your flag in your code.

◆ Variation 1   Name (optional)   Description (optional)

true			---

◆ Variation 2   Name (optional)   Description (optional)

false			---

Default variations ❶

ON	◆ true	⌄		OFF	◆ false	⌄

☐ This is a permanent flag ❶

`SAVE FLAG`

Figure 9.10 – Creating a new feature flag (the remaining portion of options)

The remaining options to create a feature flag are all concerned with the variations that this flag can provide. Four options are available for the variations that will be returned:

- **Boolean**: This is the most common type of feature flag and the one we have looked at most throughout this book. It offers two states for a flag as it either serves **true** or **false**. This is used for **A/B tests**, **switches**, and when looking to roll out a new implementation to replace an existing one.

  The value of the variation itself cannot be set as it is either true or false, but optional **Name** and **Description** attributes can be added. Setting these attributes can help provide clarity on the flag and what the two variations will do within the application, so it is a good idea to set them. For larger teams and more complex applications, there is more value in having extra information on flags to provide additional contexts, such as setting the **Name** and **Description** fields.

- **String**: A string flag variation allows a string to be returned rather than a **Boolean** value. This might not seem all that different from a Boolean flag, but it is possible to use more than two variations with this type of flag, which opens up the opportunity for **multi-variant testing**.

  The variations can, again, have an optional **Name** and **Description** but require a string value to be provided. What strings are to be used for this type of variation is open – it depends on how the team wants to work and how the flag is to be used within the application. Often, more semantic naming is useful, so it could be that the variation's string value and **Name** are the same to make reading the code simpler, and also less likely that the flag is implemented or configured incorrectly.

- **Number**: The number variation type is similar to a string type, but only integer values can be used. In some cases, it might be preferable to use the number variation type for a feature flag, since an enum can be used within the code base to map the numeric value of the flag to something more meaningful within the application.

  The number type can also be useful if you want to iterate quickly with an implementation of a feature and are running several experiments. Numbers help keep track of the experiment being run that corresponds to a log of what has been changed within the implementation. Trying to use a string variation can be done just as easily, but trying to name each experiment might be unnecessary.

- **JSON**: The JSON type of variation is the most complex to set up but works just like the string and number ones. Each variation expects a JSON object to be added, and this is what is returned when the flag is evaluated. This type of flag can be especially useful for quick experimentation, where changes only need to be made within LaunchDarkly for several aspects of an application to change.

One of the key differences with using a JSON variation is that the application needs to work a little differently from the examples I have shared, since content and/or configuration is coming back from LaunchDarkly. In the examples provided, we saw how LaunchDarkly can be used to execute and render different parts of an application, rather than to provide the differences itself. Using JSON variations requires applications to be architected to rely on this data object.

If a JSON variation flag has been configured to be a permanent flag, then it could be that LaunchDarkly is relied upon within an application to manage configuration differences across different customer segments. For example, the different configurations could be used for entitlement-based scenarios to serve specific JSON objects for customers in different countries, or even for distinct types of customers. As I have cautioned in the past, this type of application configuration can be dangerous to have permanently within LaunchDarkly. If there are any service issues with LaunchDarkly, then your application's configuration can't be retrieved and customers won't get the experience they expect.

For all variation types, there is the ability to set the default **ON** and **OFF** states of a flag. When configuring a flag that is not using the **Boolean** variations, it is worth checking that the defaulted variations are the ones expected to be **ON** and **OFF**. Usually, you will want the **OFF** state to be the control or current implementation. The default **ON** state can be any of the variations. Once the flag has been implemented within the code base and deployed to production, the flag will start being evaluated. Having the security of knowing that the default **OFF** state will not negatively impact customers is always important.

This also relates to the pattern that we covered in *Chapter 3, Basics of LaunchDarkly and Feature Management*, in the *Improving your first feature flag* section, around always implementing the default/off state of a flag in such a way as to not change the existing implementation. Ensuring that the configuration of a flag mirrors this approach is important in ensuring that a key aspect of feature management is achieved, which is reducing the risk of performing deployments that degrade production.

The final option that's available when creating a flag is to make the flag a permanent one. This topic was covered in detail in *Chapter 6, Switches*. Permanent flags have valuable use cases, especially when teams can turn non-essential parts of a website off to mitigate any issues within the system. When setting a flag to be permanent in LaunchDarkly, it can help to manage the flags within a project so that the tool does not prompt to remove the flag from the code base, or to have a dashboard setup that only shows temporary flags.

What we have not covered within this book yet is deleting feature flags. Before a flag can be deleted, it must be archived, so we will look at that before the end of this section. The final part of the LaunchDarkly feature flag dashboard to explore is the hamburger menu on the right-hand side of the grid, as shown in the following screenshot:

Figure 9.11 – The hamburger menu on the feature flag dashboard

There are two pieces of functionality here: the ability to compare flags' settings across environments and the ability to view archived flags. We will look at both next.

Comparing flags

When working with feature flags on a test environment, many properties must be configured so that targeting a flag works as needed. Once a feature has been deployed to production and the development is completed, it is valuable to be able to see how the feature flag has been set up across the test and production environments. By being able to compare feature flags, LaunchDarkly helps highlight any discrepancies between a flag across multiple environments:

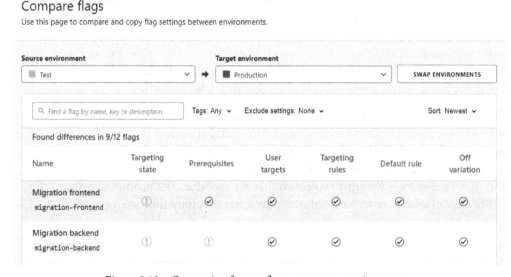

Figure 9.12 – Comparing feature flags across two environments

When comparing flags, the first thing to do is select which environments to compare the flags across. Throughout this book, we have only used the test environment, so we should expect that, when comparing the flags, there will be discrepancies with the setup between test and production. These issues are shown with a warning icon. When clicking on a flag, a panel is presented that shows what these differences are:

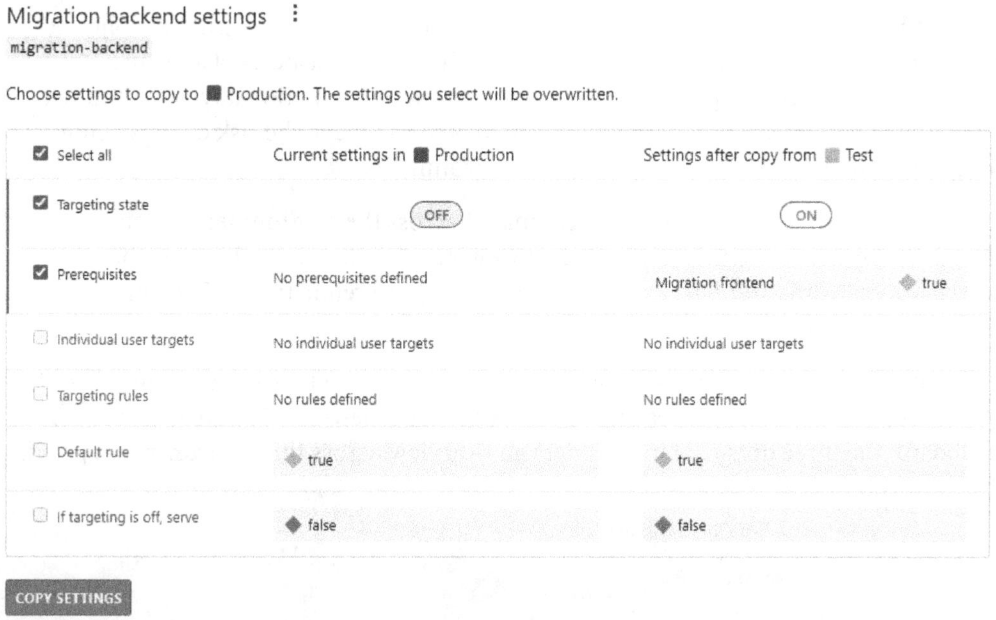

Migration backend settings ⋮
`migration-backend`

Choose settings to copy to ■ Production. The settings you select will be overwritten.

☑ Select all	Current settings in ■ Production	Settings after copy from ▨ Test
☑ Targeting state	OFF	ON
☑ Prerequisites	No prerequisites defined	Migration frontend ◆ true
☐ Individual user targets	No individual user targets	No individual user targets
☐ Targeting rules	No rules defined	No rules defined
☐ Default rule	◆ true	◆ true
☐ If targeting is off, serve	◆ false	◆ false

COPY SETTINGS

Figure 9.13 – Comparing a flag across two environments

When viewing the differences between two environments for a particular flag, the screen not only highlights where the two differ but also allows for the configuration to be easily copied from one environment to the other. The order of the environments selected when first viewing the comparison dashboard determines which environment the settings would be copied from, with the settings copied from the one selected first.

As shown in the preceding screenshot, there are two differences with this flag's settings across the environment. The first is for **Targeting state**, which is currently **OFF** for the production environment; if the test value were to be copied over, then it would be made **ON**. The second is the **Prerequisites** configuration, which on the test environment is set to the **migration frontend** flag. This needs to be set to `true` before this flag will be evaluated, whereas in the production environment, there are no prerequisites at all.

Choosing which elements of the configuration should be copied can easily be achieved by using the tick boxes on the left of the panel. In the preceding screenshot, both the **targeting state** and **prerequisite** settings have been selected to be copied over. Once the necessary aspects of a flag have been selected to be copied, by clicking the **COPY SETTINGS** button, the configuration will be copied from one environment to the other.

By using this experience to copy the settings, there can be a high level of confidence that what has been configured, tested, and signed off on a test environment can be replicated exactly to production. When we consider some of the valuable aspects of a feature, safely delivering features to production is a key one, and this functionality is yet another way in which LaunchDarkly can provide a way to better manage the risk of flags being accidentally configured differently across the environments.

If there are a large number of flags to be compared across the environments, then the dashboard experience is similar to that when viewing all feature flags. It is possible to search for keywords and filter on some of the settings to refine the list to be of a manageable number.

There is another way to view the configuration of a flag across environments, and this can support more than just comparing across two. When viewing a feature flag on a dashboard, the three dots (**...**) menu offers an overview across the **Environments** panel:

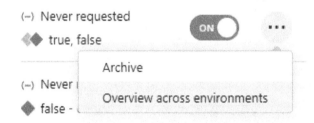

Figure 9.14 – The three dots menu of a feature flag in a dashboard

This view does not provide such a clear comparison, nor does it allow us to easily copy settings from one environment to another for a flag. However, it can be useful to view the state and configuration of a flag across all environments, as shown here:

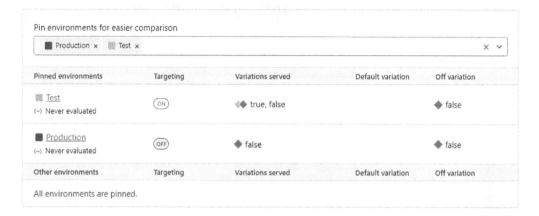

Overview across environments

Review information for **Migration backend** across all environments.

Pin environments for easier comparison

Production ×   Test ×   × ∨

Pinned environments	Targeting	Variations served	Default variation	Off variation
Test (–) Never evaluated	ON	true, false		false
Production (–) Never evaluated	OFF	false		false
Other environments	Targeting	Variations served	Default variation	Off variation

All environments are pinned.

Figure 9.15 – An overview of a flag across multiple environments

This experience shows the value of some of the configurations for a flag across the environments, rather than showing what a value would be if it were to be copied from one environment. Therefore, using this view, it might be a little easier to understand the settings of a flag. This view allows some environments to be pinned at the top; there are only two setups in this example project, so the value is limited. In a project that has many environments, it can be helpful to see a few key environments' differences. This overview experience is yet another way to ensure that feature flags have been configured in the correct way to, again, bring peace of mind while deploying code and ensuring that the right settings are being implemented in production.

The other piece of functionality within the hamburger menu on the main feature flag dashboard is the ability to view archived flags, which we will look at next.

Viewing archived flags

Before a feature flag can be deleted, it needs to be archived. To do so, we will look at the dashboard of the archived flags. This experience is familiar to that of the non-archived flags:

Figure 9.16 – The archived feature flag dashboard

The first thing to note is that there is a subtle pattern at the top of the page to indicate that the flags being shown are the archived ones. This is because nearly all the functionality and information presented on the archived dashboard is the same as the non-archived dashboard, so you could be easily looking at the wrong one. The only real difference between this dashboard and the regular ones we have already seen is that there is a **DELETE** button.

To switch back to the regular feature flags, there is an option to **View active flags** under the hamburger menu when viewing the archived flags.

The reason a flag should be archived before being deleted is to offer an element of protection from accidental deletions. Once a flag has been deleted, it is gone, and this could have a massive negative impact on an application, should this be done mistakenly. While an archived flag will no longer evaluate like a regular flag, it is at least quick to un-archive and get the application back to a good state if this was done accidentally.

Archiving a feature flag

There are a couple of ways to archive a feature flag, either through the three dots (...) menu of a single flag on the dashboard or within the settings when viewing a flag (we will look at the settings of a flag later in this chapter). Several things need to be considered when archiving a flag and LaunchDarkly helps with this, as seen here:

Archive this flag?

This will archive **Archived flag**, and the value defined in your code will be returned for all your users. Remove code references to `archived-flag` from your application before archiving.

Check flag state before archiving

Choose production environments to confirm there are no unintended effects in archiving your flag. This flag will be archived in all environments.

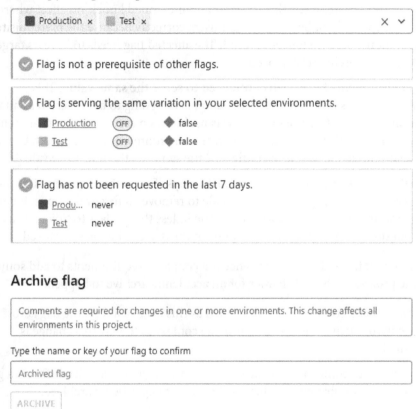

Archive flag

Comments are required for changes in one or more environments. This change affects all environments in this project.

Type the name or key of your flag to confirm

Archived flag

ARCHIVE

Figure 9.17 – Archiving a feature flag

The first thing to note with this screen is the warning, stating that all code references to a flag should be removed before the flag is archived. This is a good practice and if this is not done, then the hardcoded default value of the feature flag will be used as the flag will no longer be evaluated. In the scenario of a feature rollout, the hardcoded default behavior should result in the original implementation being executed, rather than the new one that has been rolled out to customers. Therefore, it is highly advisable to remove the feature flag's encapsulation first.

Next, the flag archive screen provides a useful overview of the configuration and usage of a flag. It is possible to specify which environments are being considered when archiving a flag, which can help when dealing with flags in a large project. There may be some test environments that would not be impacted, should the flag be archived, so they could be excluded from the overview information.

The three important pieces of information provided by the overview are as follows:

- **Whether the flag is a prerequisite**: If a feature is a prerequisite to another flag, then it cannot be archived. This prevents an unintended knock-on effect where another flag would not be able to target users correctly because the prerequisite configuration would no longer be valid. The affected flag needs to have its targeting updated before archiving can occur.

- **If all environments have been configured to serve the same value**: This overview provides certainty that production is not going to be impacted as all the environments are set up the same. This is useful when automation tests are relied upon, to prove that everything works on test environments as the expectation needs to be that production has been configured the same for the tests to be valuable.

- **If the flag has been evaluated in the past 7 days**: This piece of information shows whether the code changes have been made to remove all flag evaluations. It could be possible that work has already been done in less than 7 days to tidy up the code base, but this is a good indication to know if the flag can be safely archived.

The final element of the archiving experience is a couple of requirements to add some friction to the process so that it is harder for an accidental archive to happen:

- The first is that a comment must be added as this change affects the flag across all the environments; it is useful to provide context as to why this change is being made.

- The second is to confirm the name of the feature flag being archived. This is a good opportunity to double-check whether the correct flag is being archived.

Once the flag has been archived, it will no longer appear in the regular feature flag dashboard since that only shows the active flags. Rather, the flag will now appear in the archive flag list. An archived flag can still be looked at to view all the settings and configuration information if needed. The only real difference is that it cannot be evaluated by applications anymore.

Once a flag has been archived, it can be deleted. Deleting a flag is the next piece of functionality we will look at.

Deleting a feature flag

Deleting a feature flag once it has been archived is quite simple. Again, there are a couple of ways of doing this, either from the archived flags dashboard or within the setting of the flag itself. A popup is shown when deleting a flag:

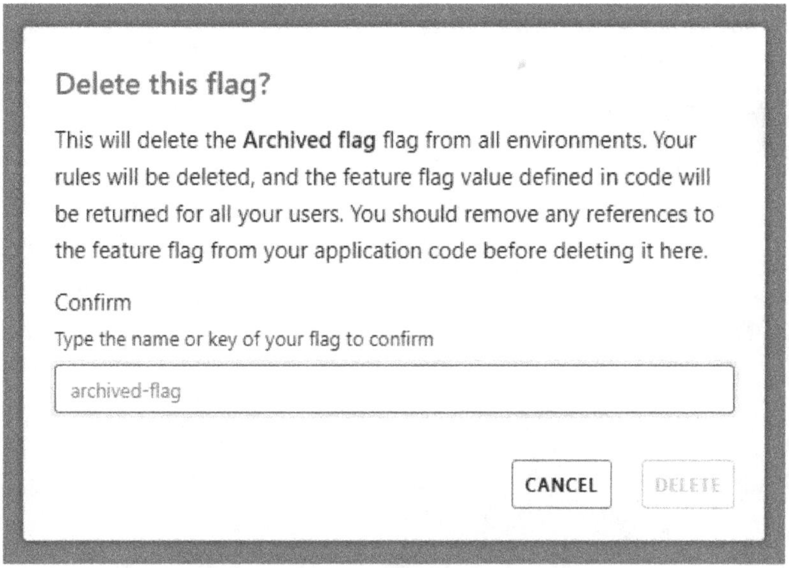

Delete this flag?

This will delete the **Archived flag** flag from all environments. Your rules will be deleted, and the feature flag value defined in code will be returned for all your users. You should remove any references to the feature flag from your application code before deleting it here.

Confirm

Type the name or key of your flag to confirm

archived-flag

CANCEL   DELETE

Figure 9.18 – Deleting a feature flag

Deleting a flag is a straightforward process since you only need to confirm the name or key of a flag. This might seem too easy to do but this is because all the necessary checks were done when the flag was archived. This means that once archived, a flag can be safely deleted. Again, this is an opportunity to ensure that the correct flag is being deleted. Unlike archiving a flag, this cannot be undone and once the flag is deleted, its settings and configuration are no longer viewable. Once deleted, the flag will not appear in any dashboards.

In this section, we looked at how to make the most of the features available to manage feature flags within a project, how to create a flag, and how to archive and then delete a flag. In the remaining sections of this chapter, we will look at the functionality available on a feature flag to make the most of all the features LaunchDarkly provides. Next, we will look at targeting and variations.

Working with targeting and variations

The key to feature management and where LaunchDarkly really shines is in its ability to target feature flags' variations to specific users. This is done within the **Targeting** section of the **Feature flags** page. There are four key aspects to **Targeting**, and they are **Prerequisites**, **Target Individual users**, **Target users who match these rules**, and **Default rule**. These are shown in the following screenshot:

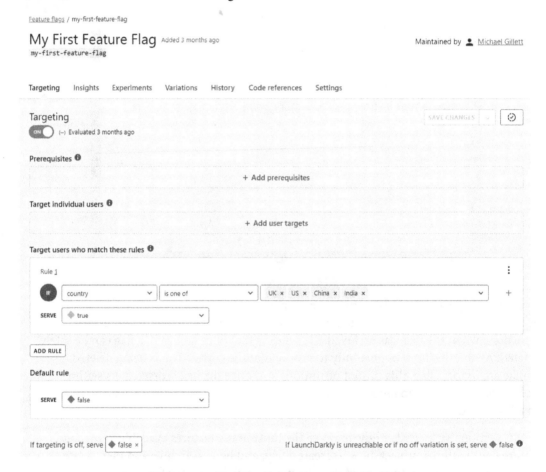

Figure 9.19 – The Targeting section of a feature flag

Much of this functionality has already been discussed in detail in previous chapters: targeting individual users or segments and targeting users who match a rule and default rule in *Chapter 3, Basics of LaunchDarkly and Feature Management*, and *Chapter 4, Percentage and Ring Rollouts*, respectively, and **Prerequisites** in *Chapter 8, Migrations and Testing Your Infrastructure*.

Next, will provide a simple overview of this section of the feature flag since more detailed explanations and use cases were provided in the aforementioned chapters. One thing to consider is that the order of the targeting configuration is important. Moving down the page, we will see broader and broader targeting rules; if a feature flag has not met the criteria of the current rule, then the evaluation will move on to the next rule. For example, the first check is for any prerequisite flag evaluations. If none have been configured, then it will look for specific users and move on to targeting based on any configured rules before ultimately using the **Default rule**. The higher up the page a setting is, the sooner within the evaluation it will have an effect. Targeting begins with more specific rules and moves to a broad default behavior by the end of the evaluation.

It is worth noting that only when targeting is enabled will any of the following configurations for a flag be applied. When targeting is **OFF**, a single variation can be specified to be returned. It is also worth considering that targeting rules are unique for a flag in each environment. When looking to set a certain targeting configuration from one environment to another, it is useful to use the **Compare flags** tool, which we looked at in the *Comparing flags* section of this chapter.

The **Prerequisites** configuration allows flags' evaluations to link to another flag. This allows complex pieces of work to have multiple feature flags within the code base, which helps you control the implementation and rollout. However, you can have a single flag that determines the variation to be served to a user. This can be useful in large migrations or where complex user journeys are being experimented with.

Targeting individual users and segments is often done for testing and sign-off purposes, where key people should receive a certain variation of the flag. By explicitly setting a user to receive a variation, whenever the flag is evaluated for that customer, they will always get that variation.

Targeting a flag variation for a configured rule is very important when it comes to rolling out and experimenting with features. Several default attributes are provided by LaunchDarkly that can be used for targeting a variation. But it is also easy to add custom attributes to enable businesses to add their own custom data types to be used to serve different outcomes from a flag.

Default rule is self-explanatory and is the variation that is to be served, should a user have not already received an evaluation from targeting specific users or through any of the rules. This default variation is likely to always be the current implementation of a feature in rollouts and experimentations. It is also likely to be the same as the default value to be served when targeting is off, but there may be instances where this is not the case.

At the bottom of the section is the default variation to be served when targeting is off. As a best practice, this should be the false variation or whichever variation is being used as the control/current implementation. This ensures that the current functionality is given to customers before anything is configured on the flag. By not following the default pattern, it is likely that the implementation of the feature flag's encapsulation and the flag's setup will not be aligned, and that production will be impacted once the deployment containing the flag is completed.

At the bottom right of the screen is the information that's served from using **code references**, which will be explored in further detail later in this chapter.

Once a flag has been set up and the targeting has been configured, you may need to change the variations. This is likely to occur when you're working with experiments in a highly iterative manner. The **Variations** section provides an overview of this:

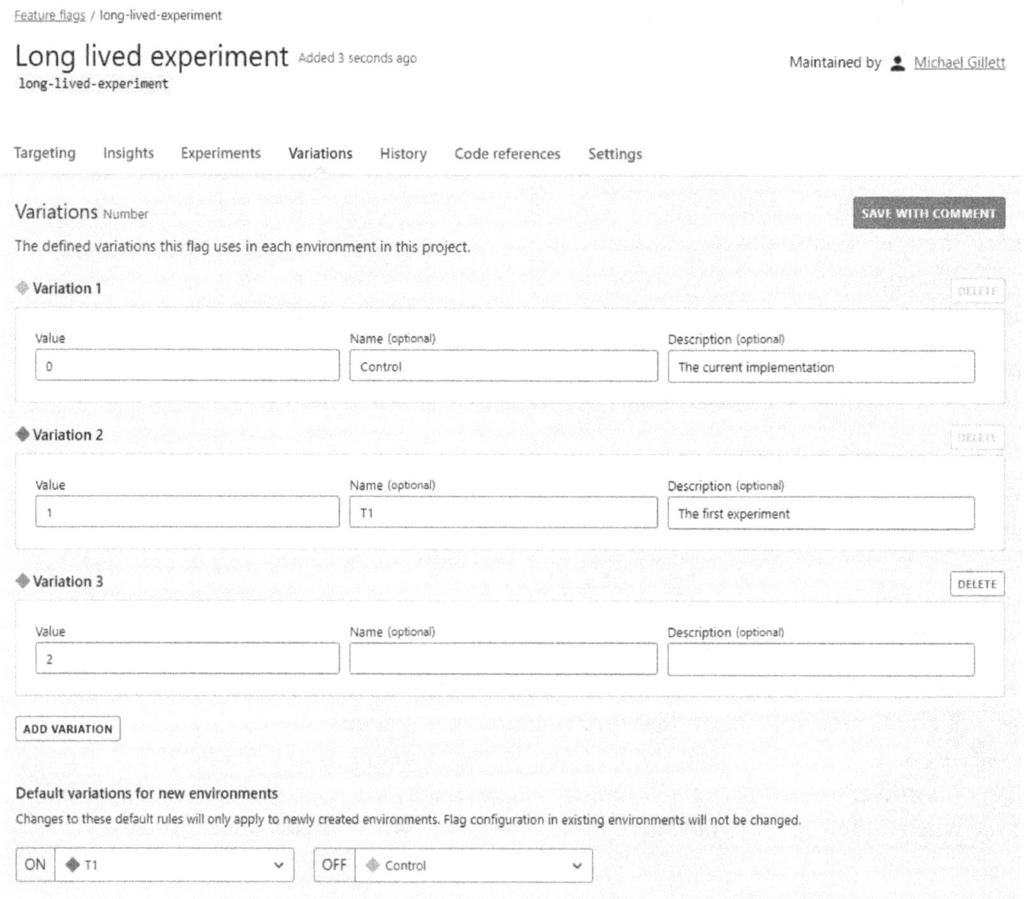

Figure 9.20 – The Variations section of a feature flag

In this example, the feature flag is using numeric variations and could be considered a long-lived experiment, where there might be a need to run concurrent and different implementations in production to see which one is performing better. With this in mind, there could be a need to add more variations over time, as seen with the addition of **Variation 3**, which doesn't have a name or description. In this scenario, those details can be added to this screen and the default variations setting can also be changed. For example, the **OFF** variation would remain the **Control**; however, the new variation, **Variation 3**, could now be set as the default **ON** one.

As variations become obsolete, they can be deleted from the feature flag from this screen.

One thing to note is that if a feature flag is part of an experiment, then the values of the variation cannot be changed while the experiment is recording data.

One thing that has not been looked at in this book is what kind of data LaunchDarkly provides for seeing how the different variations are being evaluated. This includes seeing the numbers of each variation being served and when changes were made to the flags that results in significant changes in what was evaluated. This is what we will look at next.

Discovering insights and history

The **Insights** section of a feature flag shows what variations have been served by the flag over time. This is presented in a graph with several time frames available. By default, it shows the last 60 days:

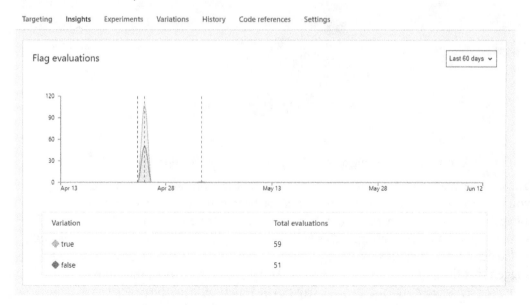

Figure 9.21 – The Insights section of a feature flag

The chart shows the number of times each variation was evaluated over the time frame. In this example, most of the requests to evaluate the feature flag were on the same day and the targeting was configured to return a 50/50 split for most users, with some requests coming from a specific user who should always receive the **true** variant.

Overlaying the chart are dotted lines that indicate the times when the configuration of the flag was changed. By being able to see both the variant evaluations and when changes were made, this chart can prove useful when analyzing how the flag is performing and what changes have impacted the variants being served. When hovering over one of the dotted lines, more information is revealed, including the number of evaluations on that day and a summary of what changes were made. Clicking the line will reveal all the changes that were made on that day:

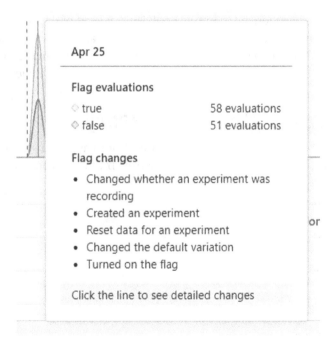

Figure 9.22 – Hovering over the Insights chart reveals additional information

There are several scenarios where using the **Insights** section can be useful, including when you want to ensure that a percentage rollout has been configured correctly and when a flag has been completely rolled out. In both use cases, the chart can quickly indicate if something has not been configured as expected. For example, if a flag is meant to be 50/50 rolled out but the insights show that variants are not being evenly split, then this would suggest that the targeting needs more work. Equally, if there is more than one variant being served when it was thought the flag can be removed from the code, it would show that some form of targeting is still incorrectly configured.

While LaunchDarkly offers other ways to understand if a flag is ready to be removed and archived, the **Insights** chart can be used to quickly see this too. This chart also provides a useful timeline to help us understand the changes that have been made to the flag over time. There is, however, another section that shows the history in far greater detail, which we will look at now:

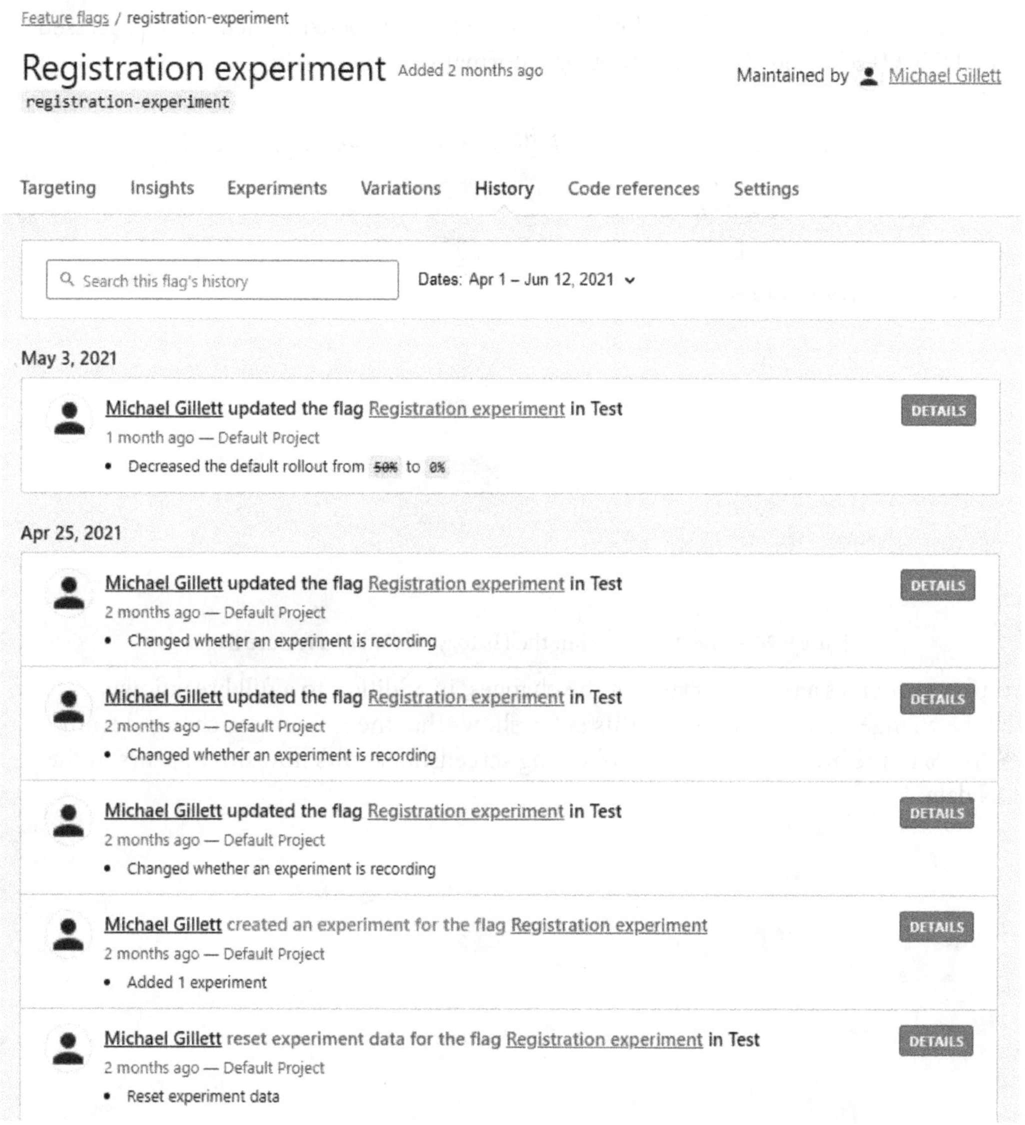

Feature flags / registration-experiment

Registration experiment Added 2 months ago

registration-experiment

Maintained by 👤 Michael Gillett

Targeting   Insights   Experiments   Variations   **History**   Code references   Settings

🔍 Search this flag's history   Dates: Apr 1 – Jun 12, 2021 ⌄

May 3, 2021

👤 **Michael Gillett** updated the flag Registration experiment in Test
   1 month ago — Default Project
   • Decreased the default rollout from 50% to 0%   DETAILS

Apr 25, 2021

👤 **Michael Gillett** updated the flag Registration experiment in Test
   2 months ago — Default Project
   • Changed whether an experiment is recording   DETAILS

👤 **Michael Gillett** updated the flag Registration experiment in Test
   2 months ago — Default Project
   • Changed whether an experiment is recording   DETAILS

👤 **Michael Gillett** updated the flag Registration experiment in Test
   2 months ago — Default Project
   • Changed whether an experiment is recording   DETAILS

👤 **Michael Gillett** created an experiment for the flag Registration experiment
   2 months ago — Default Project
   • Added 1 experiment   DETAILS

👤 **Michael Gillett** reset experiment data for the flag Registration experiment in Test
   2 months ago — Default Project
   • Reset experiment data   DETAILS

Figure 9.23 – The History section of a feature flag

The **History** section provides all the information required for the changes that have been saved on this feature flag since it was created. This can be extremely useful for auditing purposes, to help us understand when changes were made and, should comments be added when saving a change, why the change was made. It is also useful should there be a need to return a flag to a previous state as it is easy to see what has been changed since that version.

Lots of details about each entry can be found in the **History** panel, which can be accessed via the **DETAILS** button. This will reveal an information panel:

 Michael Gillett updated the flag <u>Registration experiment</u> **in Test**

May 3, 2021 1:23 PM — Default Project

Changes:

* Decreased the default rollout from ~~50%~~ to 0%

Diff

```
{
  "archived": false,
  "clientSideAvailability": {
    "usingEnvironmentId": false,
    "usingMobileKey": false
  },
```

Figure 9.24 – The details within the History section of a feature flag

This panel provides a brief overview of the changes, in addition to a full breakdown of all the changes in a JSON object. This entry shows that the rollout was changed from 50% to 0% in the brief overview. The following screenshot shows how this appears in the JSON data:

```
"test": {
  "archived": false,
  "fallthrough": {
    "rollout": {
      "variations": [
        {
          "variation": 0,
          "weight": 50000
          "weight": 0
        },
        {
          "variation": 1,
          "weight": 50000
          "weight": 100000
        }
      ]
    }
  }
},
```

Figure 9.25 – The JSON information provided for each entry in the flag's history

This granular information is particularly useful when it comes to understanding how a flag was configured and what exactly has been changed within it. It is not all that often that the history of a flag needs to be accessed, but knowing there is such a detailed record of the changes does provide reassurances for any changes to be fully understood and audited.

Between the **Insights** and **History** sections, the impact of changes to the flag can be observed on the flag's performance and reverted, should the changes not result in the desired outcome. Having this data readily available is important and complements the functionality and data that's surfaced in the **Experiments** section of the flag, which we will look at next.

Learning about experimentation

The **Experimentation** section of a feature flag was covered in detail in *Chapter 5, Experimentation*, so it is best to read that chapter to understand how best to use this functionality, along with **Metrics**. However, there are a couple of additional pieces of functionality that were not covered in that chapter that we will cover next.

When working with the experimentation functionality, it is possible that testing can distort the results of the test. Usually, the only data that should be collected and analyzed is that of the percentage or ring rollout of a flag to customers. When testers or stakeholders are validating or signing off a feature, their usage of an experiment should not impact the data. Normally, the sign-off will be performed before an experiment is presented to any customers. However, testing might continue in production (even once live) to ensure that all the variations of an experiment work. In that case, specific users or targeting rules might be set up to have the flag serve particular values. In the targeting section of a feature flag, individual users can be targeted, as shown here:

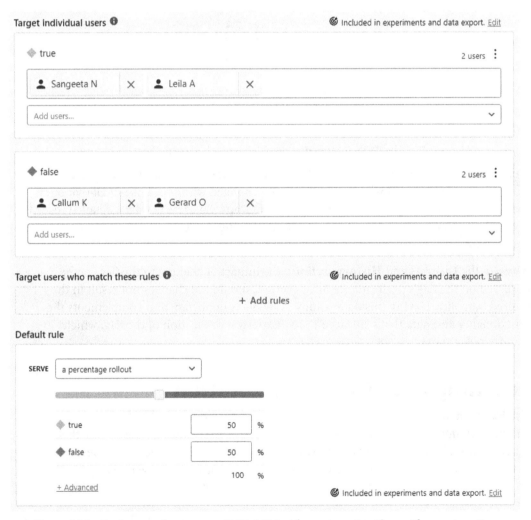

Figure 9.26 – Setting up a flag to run a 50/50 A/B test for customers with specific user targeting

Here, we can see how both a 50/50 rollout can be configured, alongside targeting specific users to get either outcome. Within the experiment itself, we do not want the results of the individually targeted users to impact the results. In the **Experimentation** section, it is possible to see and change the targeting rules that are being used in the experiment. By default, it will record all the targeting rules. For this flag, we have set up two rules:

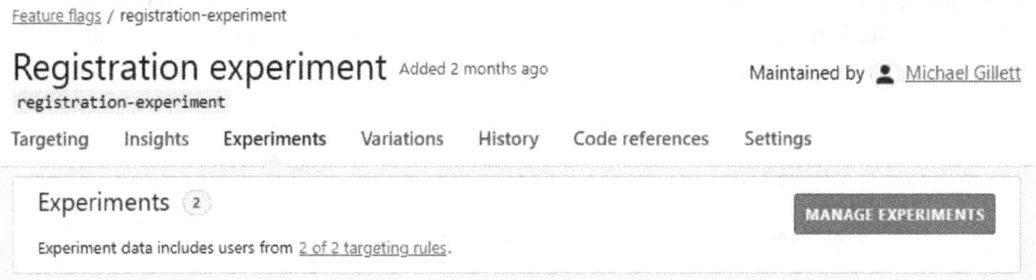

Feature flags / registration-experiment

Registration experiment Added 2 months ago
registration-experiment

Maintained by 👤 Michael Gillett

Targeting   Insights   **Experiments**   Variations   History   Code references   Settings

Experiments ②

MANAGE EXPERIMENTS

Experiment data includes users from 2 of 2 targeting rules.

Figure 9.27 – The Experiments section shows how many targeting
rules are being included in the experiment data

LaunchDarkly allows us to change the targeting rules being used within the results of the experiment. This can be accessed via the **2 of 2 targeting rules** link, which opens a new panel, as follows:

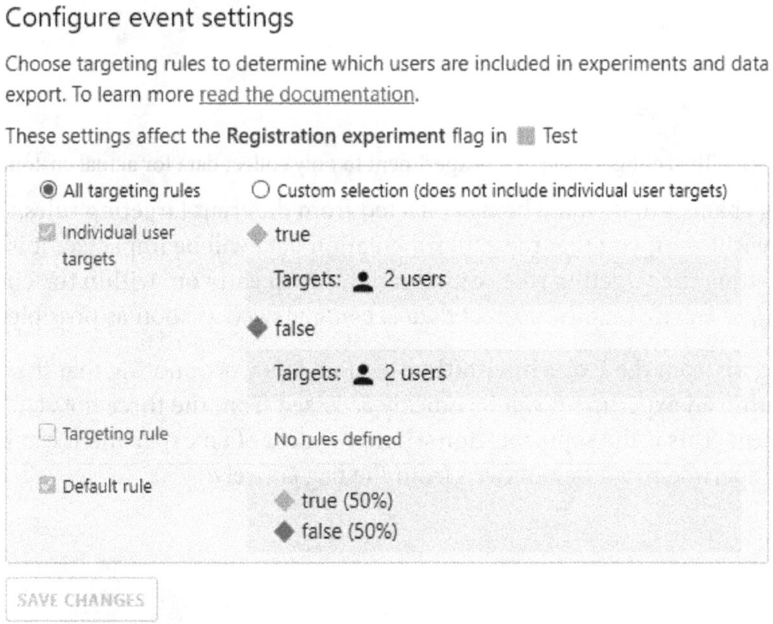

Configure event settings

Choose targeting rules to determine which users are included in experiments and data export. To learn more read the documentation.

These settings affect the **Registration experiment** flag in ▦ Test

◉ All targeting rules   ○ Custom selection (does not include individual user targets)

☑ Individual user   ◆ true
  targets
                            Targets: 👤 2 users

                         ◆ false

                            Targets: 👤 2 users

☐ Targeting rule   No rules defined

☑ Default rule
                         ◆ true (50%)
                         ◆ false (50%)

SAVE CHANGES

Figure 9.28 – The targeting rule configuration panel in the Experiments section

By default, all the targeting rules are applied; however, a customer selection can be configured instead. In this scenario, we are only interested in the **Default rule** property being used as the data being collected from the metric for this experiment. The users within this rule are going to be customers, so they will provide the data of value to help with the success or failure of the experiment. Users within the **Individual user targets** rule are going to be people within a team who were given access to test and sign off the feature; their data is not useful for the success of the experiment. The ideal configuration for data being gathered would look like this:

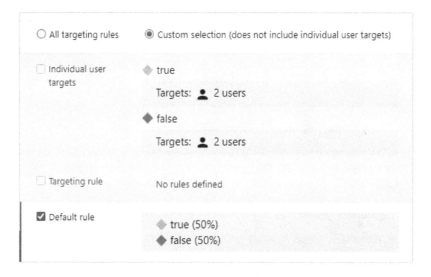

Figure 9.29 – The configuration of an experiment to only collect data for actual customers

When making changes to the data being collected from different targeting rules, LaunchDarkly will warn you that the experimentation data will be impacted. It is worth considering getting the targeting rule configuration set up early on, within the course of an experiment, to ensure that the correct data is being viewed as soon as possible.

Before moving on from the **Experimentation** section, it is worth noting that the metric being used within an experiment can be quickly accessed from the three dots (…) menu of an experiment. This is the same location where the data of an experiment can be reset, or where the experiment can be removed from the flag entirely:

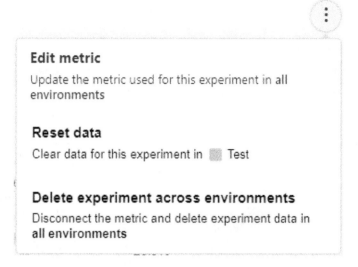

Figure 9.30 – The menu for updating or removing an experiment

We looked at resetting experiment data in detail in *Chapter 5, Experimentation*. It is important to remember that experiments are set up across all environments when they're added to a flag in a single environment. The data of an experiment is unique to each environment. Therefore, data can be reset for one environment, but deleting an experiment applies to all environments.

The other useful piece of functionality that has not been covered yet regarding experiments is that the time frame can be configured to view the outcome of an experiment across a given time frame. This can be useful if changes have been made to the targeting of a flag or if improvements have been made to a feature. Usually, it is better to run each iteration of a feature as a separate experiment to truly understand the success or failure of a version of a feature. However, sometimes, iterations might be made to a feature during a live experiment, so viewing the results during a specific date range is important.

This can be done on the result view:

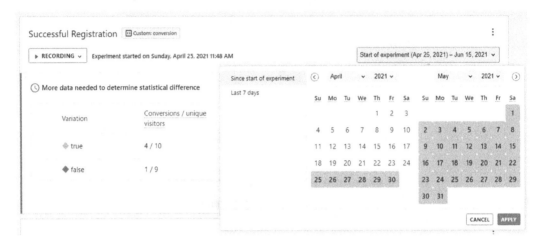

Figure 9.31 – Selecting a time frame to view the experiment data

Having such granular control over what data is used within an experiment, and then being able to refine the date range in the **Experimentation** section, is a very useful way to truly understand the success of an experiment. This functionality can empower teams to become hypothesis-driven in their approach to building software. This enables them to prove the best experience and functionality for your customers. Seeing what variation was most successful during defined periods provides the opportunity to fully understand how customers interact with products.

Next, we will touch on how the feature flag can be tidied up and what further settings are available to manage flags.

Using code references

The **Code references** section is only a small component of a feature flag and was detailed in *Chapter 8, Migrations and Testing Your Infrastructure*. Please refer to that chapter to learn how to use code references. However, there is one thing to note that was not covered in that chapter.

Figure 9.19 shows the default variation to be served at the bottom right, where it says **If LaunchDarkly is unreachable or if no off variation is set, serve false**. LaunchDarkly's code reference tool has detected where the flag is implemented within the code and reads the value that is hardcoded as the default value. This is useful information as a flag could be configured within LaunchDarkly to return `true` for all targeting rules and even when targeting is disabled. However, if the service is unreachable and the default value is hardcoded as `false`, then that is the value that the application will work with.

In this scenario, it is a good practice to remove feature flag implementations quickly following a successful rollout or experiment to ensure that the code base only has the desired implementation within it. Using code references in this way can help you identify where a flag's hardcoded default will differ from the defaults configured in LaunchDarkly. The last section of a feature flag is **Settings**, which we will explore next.

Exploring Settings

Settings contains several properties and configuration options that are available through other areas of LaunchDarkly, but it does offer some unique functionality too. The first setting that can be changed is the data export configuration, which is what we previously explored within the **Experimentation** section in terms of selecting which targeting rules should be used when evaluating a feature flag experiment. Most of the functionality is the same as within the **Experimentation** section. However, it is possible to disable the ability to send detailed event data to the export locations. When using the data for experimentation, detailed data needs to be sent, and this cannot be turned off.

The second piece of functionality within **Settings** is **Triggers**. **Triggers** is a feature that can only be found in enterprise-level subscriptions, so it is not available on every plan. Triggers provide the opportunity for other services and systems to control the targeting of a flag. There are several supported services available when setting up a trigger, but a generic webhook trigger can also be created. Each trigger generates a unique URL for that flag on the environment were it was created, and it is possible to have several triggers on a flag. With this in mind, triggers need to be created across every environment for a flag where they will be used.

There are several opportunities available to use these triggers, such as automatically turning off feature flag targeting when an automated alert has been raised about the health of the system. This results in a quicker response and does not need manual intervention. Once the alert has been resolved, the flag's targeting can be re-enabled. This allows the system to return to its expected state. A practical example of this could be to automatically turn off a **kill switch** during any instability in the production environment. It is often only in production where this automation is required and where it offers the most value. This common use case is why a number of the supported third-party integrations are with **Application Performance Monitoring** (**APM**) tools such as DataDog, Dynatrace, and New Relic.

The next area of the **Settings** section contains a number of the fields and options that are available when creating a flag. This includes **Name, Description**, and **Tags**. It is possible to change the **Maintainer** property of the flag, which is set to the account that created the flag by default. The **Key** property of the flag is also visible here, although it cannot be changed. It is also possible to change the configuration of the client-side SDK availability here. As a feature is developed and rolled out, there could well be a need to change the information and setup from when the flag was first created.

The next part of the **Settings** section is **Custom Properties**. This is useful if there is a need to integrate another system with LaunchDarkly and link feature flags to elements within the other tool. A custom property could be created that contains metadata for a flag to link it to a third-party system. For example, integrating flags with Jira or other ticket management systems can be useful and in this case, the number of a Jira issue could be added to a flag as a custom property. It is possible to add any type of custom property, so bespoke tools can be made to add value to business processes if needed.

The remaining sections of **Settings** are all concerned with managing the life cycle of feature flags. The first is that it is possible to clone a feature flag that includes all the targeting rules and targeting states across all the environments to a new flag. All that needs to be provided for this is a **Name** and a **Key**. There is the option to add a **Description** and **Tags**, and this experience is like creating a new flag. The other flag management functionality is to archive the flag. This offers the same experience as shown in *Figure 9.18*.

Once a flag has been archived, the **Settings** section changes slightly. The **Archive** functionality is replaced by the option to **Restore** the flag. In addition to this, there is the option to completely delete the flag. Again, this experience is just as we saw earlier in this chapter for deleting a feature flag.

The **Settings** section offers a consolidated location to manage a feature flag, even though much of this functionality appears throughout other parts of LaunchDarkly too. With the ability to set up triggers and custom properties, it is possible to extend the normal experience of working with flags to provide more automation and integration with other systems to build on top of what LaunchDarkly provides.

Summary

From reading this chapter, you should have gained an insight into the full functionality available for working with and managing feature flags. Before this chapter, this book had only detailed parts of LaunchDarkly relevant to the topic being explored at that time, whereas this chapter has covered everything that LaunchDarkly provides. This should enable you to make the most of the tool.

You should now know how to work with feature flags throughout their entire life cycle, from creating flags to archiving them, to deleting them completely. The use of filtering and refining the views and dashboards when working with a large number of concurrent flags will help you keep using LaunchDarkly efficiently.

Being able to extract the most value from the data provided within a feature flag is important, and you should now be able to use both the **Insights** and **Experimentation** sections to see how a flag is performing, either by ensuring that the targeting has been configured correctly or that one variant is performing better than another. In addition to this, being able to use the **History** section to see the impact that certain changes have had can be highly effective at understanding or reverting the changes.

By taking a quick look at the **Code references** and **Settings** sections, you should know how to manage a flag while developing a feature and updating the information attached to it, and feel confident in tidying it up at the end.

While a feature flag is a simple concept, there is so much that LaunchDarkly provides to maximize its functionality that it is important to understand what is possible. In the remaining chapters of this book, we will look at other key areas in similar detail to empower you and your teams to make the most of LaunchDarkly.

In the next chapter, we will look at **users** and **segments**. Effective targeting is the next step in being able to make the most of the LaunchDarkly tool now that we have looked at feature flags in detail. From understanding how to implement the User object within your applications to how to group users, this will enable feature flags to be efficiently and effectively targeted at your customers.

10
Users and Segments

Since we have looked at feature flags in both the broad sense and then in detail in *Chapter 3*, *Basics of LaunchDarkly and Feature Management*, and *Chapter 9*, *Feature Flag Management in Depth*, respectively, it is now time to explore **users** and **segments** in detail. Being able to understand the User object, as well as the functionality that LaunchDarkly provides to target and manage users and groups, is key to being able to make the most of feature management.

While the implementation of a feature flag is simple, with it effectively being an if statement, the real power comes from how you work with users and the targeting within LaunchDarkly. This chapter will give you all the information you need to work confidently with users and maximize the targeting, rollout, and experimentation functionality.

We will look at how to use the LaunchDarkly **Software Development Kit** (**SDK**) to set up the User object within an application, as well as some of the optional parameters that can be set and how to provide custom attributes to the User object. This will ensure you can target users as needed and allows you to work with different types of customers and sessions. Within LaunchDarkly itself, we will look at how **users** and **segments** can be managed. We will also take a look at some of the ways they can be used in targeting that have not been covered in other chapters.

In this chapter, we will cover the following topics:

- Understanding the user

- Learning how to use segments

- How to manage users and segments within LaunchDarkly

- Exploring targeting with users and segments

By the end of this chapter, you will be equipped to implement the User object within your application and make full use of the opportunities available for targeting features. You will be able to understand when and how to target features to users and/or segments, as well as how to manage both anonymous and known customers within LaunchDarkly. This information will allow you to use LaunchDarkly for both technical and business requirements as features can be rolled out, toggled, and experimented with for specific customers or user groups.

Technical requirements

This chapter contains code samples showing how to set up and configure the User object. The examples won't show a fully working application; instead, they will show how to work with the User object. If you want to implement these samples, you can try them out using the example application from *Chapter 3*, *Basics of LaunchDarkly and Feature Management*.

The code samples will be written in C# and will make use of the .NET LaunchDarkly SDK. This is the same setup that was used in the examples from *Chapter 3*, *Basics of LaunchDarkly and Feature Management*. The code samples will differ for different platforms and SDKs, so you will need to refer to the LaunchDarkly user configuration docs (https://docs.launchdarkly.com/sdk/features/user-config) to implement it within your application.

Understanding the user

Throughout this book, we have explored various aspects of the user and how to implement the User object within your applications. It is important to understand that the user is as crucial to feature management as the feature flag itself. Without the ability to know who the user is and their details, it is very difficult to target features at specific customers or percentages or rings of customers. We have already discussed using **ring** and **percentage rollouts** to target users in detail in *Chapter 4*, *Percentage and Ring Rollouts*.

The concept of a user extends further to not necessarily dealing with actual people or customers but a request or a session. This allows systems to communicate without an actual customer making a request. With this in mind, the User object that LaunchDarkly uses can be considered as more than just a person. It can be considered as an object in which metadata can be provided to enable or disable features. For example, the version number of a backend system could be provided when creating a User object in an application so that targeting can be carried out on this property. In a frontend scenario, the version of an application might not be very relevant, but when you're working between systems, it can be an effective way to target features and roll out new implementations.

Chapter 3, *Basics of LaunchDarkly and Feature Management*, detailed how to implement the LaunchDarkly SDK and the User object. The examples provided in that chapter offer a good starting point, but some properties weren't fully explored within that chapter. To fully understand the User object, and how it can be considered as a session, we will look at some of the additional properties that are available when working with a User object.

The User object

In *Chapter 3*, *Basics of LaunchDarkly and Feature Management*, the User object was being initialized with the following implementation:

```
LaunchDarkly.Client.User.Builder(guid)
                            .Country(country)
                            .Anonymous(true)
                            .Build();
```

Within this implementation, a few things are happening. The User object gets built as the final piece of the series of methods. This allows several optional configurations to be set when the object is created. The only property that must be provided is the ID or **key** of the user, which in this example is being provided as the guid value. This identifier can be anything unique to this user or session. In this example, the user has logged out, so a consistent guid value is used for all the page loads of this person's session. This is the best way to identify them. Once a user has logged in, it is possible to use a username or other unique identifier of their account. LaunchDarkly recommends that the best key to use is a hash, if possible, to uniquely identify the user.

Since our example is that of a customer who has not logged in, the `Anonymous` method is used so that within LaunchDarkly, this user behaves slightly differently. They won't appear within the system so that they can be managed like a known user will be. This is because the expectation with anonymous users is that they are short-lived sessions and not an account that would repeatedly return to the application, or at least not in a manner where that same customer can be identified. LaunchDarkly works best when you know who the user is and can identify them across multiple sessions, systems, and products.

You might find that the anonymous user is often only used on the acquisition parts of your product, such as within home pages, landing pages, and the login and registration journeys. Experiments and rollouts in such a case are likely to be generic and will use **percentage rollouts** to deliver new functionality to customers. Once a customer has logged in, **ring rollouts** can be used when users are grouped together due to their behavior or type of customer.

There might be some scenarios where a user will be identified as an anonymous user and then log in later, thus becoming an identified user. This can be a tricky journey to implement technically as an evaluation will be performed against the anonymous users who, once logged in, could experience a different variation. This happens when a known user has already had a feature flag evaluated for them, and LaunchDarkly will remember that outcome.

To resolve this, LaunchDarkly recommends providing an additional attribute to be used as the targeting attribute instead of the default **Key** value within the targeting rules. This attribute would be a unique ID that is the same for both the anonymous and identified user so that they will not be seen as different users to LaunchDarkly. With both users having the same unique ID and flags being targeted against that attribute, the user would have a seamless experience. The caveat here is that should that user return to the application as an anonymous user after their first visit, the problem can return as the anonymous user could receive a variation that differs from when they log in.

The final method that's being used within the preceding code sample is `Country`, which is used to send the detected country of the user's session.

Default attributes

There are several out-of-the-box attributes that LaunchDarkly offers, and `Country` is one of them. These attributes are common ones for targeting feature flags at users or to help us understand and manage the user within LaunchDarkly. These other default attributes are as follows:

- `FirstName`
- `LastName`

- `Name` (this is used for the full name)
- `Avatar`
- `Email`
- `IPAddress`

The `Name`, `Avatar`, and `Email` attributes are usually only used to identify the user within LaunchDarkly as it is unlikely that you would look to target a feature based on those attributes. They are useful when you're looking to serve a specific value to a customer. Providing their `Name` makes it easier to find them within LaunchDarkly.

The `IPAddress` value can be used when targeting or identifying the customer. To target them, a range of addresses could be used: based on if a customer fits within that range, they would be served a particular variant.

Like the `Country` attribute, LaunchDarkly expects to be provided with the actual value as it does not have a mechanism to identify this itself; it is unopinionated when it comes to the detection. This allows the most efficient approach to working with `IPAddress` and `Country`.

When using the name attributes, there is no need to use the first, last, and full name together; only first and last need to be used for the `Name` method, like so:

```
LaunchDarkly.Client.User.Builder(guid)
                        .FirstName("Louis")
                        .LastName("CR")
                        .Email("LCR@example.com")
                        .IPAddress("127.0.0.0")
                        .Country("United Kingdom")
                        .Build();
```

With this type of implementation, it is easy to manage the users within the **Users** dashboard in LaunchDarkly, but there isn't much information available regarding how to target users. In the preceding example, you will notice that the `Anonymous` method was removed as the example is dealing with a logged-in user. The remaining examples will all be showing code as if the user was logged in and not an anonymous customer.

To target users based on the needs of the business, there is going to be an additional requirement to add custom values to the users. This can be done with custom attributes, which we will look at next.

Custom attributes

Custom attributes can be added to a `User` when initializing the object through the `Custom` method. This takes two parameters: the name of the attribute and the value to be used. LaunchDarkly is opinionated about the type of value being provided and with the .NET SDK, at least a specific LaunchDarkly variable type, `LdValue`, needs to be passed. LaunchDarkly provides its own types for many of the .NET strongly typed variables, including the following:

- `Array`
- `Boolean`
- `Dictionary`
- `Int`, `double`, `float`, and `long`
- `List`
- `String`
- `Object`

Adding a custom attribute is done in the same manner as using the built-in attributes. The following example shows how the subscription level can be passed through to LaunchDarkly. In a real-world scenario, the subscription might be using an enum rather than a string but for the purpose of this example, I am keeping it simple with a string:

```
LaunchDarkly.Client.User.Builder(guid)
                .FirstName("Raj")
                .LastName("D")
                .Custom("Subscription level",
                        LdValue.Of("Premium"))
                .Build();
```

With the use of custom attributes, it is possible to send any type of data to LaunchDarkly for the use of targeting when the `User` object is built. As we will see later in this chapter, LaunchDarkly shows both the built-in and custom attributes within the **Users** dashboard to be able to view the data being sent to the tool. This overview of customer data is useful when targeting features at customers, but at times, there can be a desire to target a feature using sensitive data that should not be visible within LaunchDarkly. Next, we will look at how to make attributes private.

Keeping attributes private

As teams become more experienced with feature management and targeting new functionality at customers, new opportunities can present themselves for experiments and rolling out features. This can lead to targeting users on more complex pieces of information, including those of a more sensitive nature. This information should not be visible within LaunchDarkly but could be useful for targeting purposes. An example of this could be the balance of a user's account, where a new feature is being tested for users with a large account balance.

The customer's balance should not be viewable within LaunchDarkly, but it could be very useful to target feature flags off of. It is also possible that some types of **Personally Identifiable Information** (**PII**) can be sent to LaunchDarkly for targeting, but that should not be visible within the tool. LaunchDarkly provides functionality to exclude any attributes when building the `User` object.

All attributes can be appended with `AsPrivateAttribute`, as shown in the following code sample:

```
LaunchDarkly.Client.User.Builder(guid)
        .FirstName("Kevin")
        .LastName("C")
        .Email("KC@example.com").AsPrivateAttribute()
        .Custom("Balance",
              LdValue.Of(12345)).AsPrivateAttribute()
        .Build();
```

In this example, both the built-in `Email` attribute and the custom `Balance` attribute are marked as private. All that will appear within LaunchDarkly will be the ID and the first and last names, while `Balance` and `Email` can be used within targeting rules.

Technically, what happens when marking an attribute as private is that a feature flag is targeted within the LaunchDarkly client within your application using this data. However, the private data is never sent to LaunchDarkly itself. The non-private data and the outcome of the targeting evaluation will be sent to be viewed within the tool. This allows for confidence in the fact that this private data never ends up on LaunchDarkly's servers when using the LaunchDarkly SDKs.

With this information, it should be possible to create LaunchDarkly users, as both actual customers and as sessions for the use of inter-system communication, with all the details to target features as needed. We will look at how a user can be managed once this information has been sent to LaunchDarkly, but before that, we will explore what segments are in LaunchDarkly.

Understanding segments

While users, within the context of LaunchDarkly, require implementation in the code base, the concept of segments exists within the tool itself. A **segment** can be considered a group of users that all fall within a certain category. This grouping is defined within LaunchDarkly and relies on the targeting rules we have already seen in several chapters, especially *Chapter 4, Percentage and Ring Rollouts*. We will see them in the next couple of sections as well.

Segments can be used to great effect when there are frequently used targeting rules or features that are regularly rolled out to certain groups of customers. Because the segments are created with the same attributes that are used within the targeting rules, some custom attributes might be used across several or all applications within your system to ensure segments can be used effectively. For example, there might a classification for a customer, such as *new customer*, *regular customer*, and *VIP*. By providing this classification of a user when initializing the User object, it is possible to have a segment for each of the three types of customers and use that for rolling out features. VIPs might be the last segment of customers that a new feature should be presented to, to ensure that the new feature works as expected for your most valuable customers.

In *Chapter 4, Percentage and Ring Rollouts*, some alternative uses of segments were discussed, including a certain type of testing where users could opt to beta test new features. This differs from the percentage rollout as these customers know they are part of a test. Again, this would require new custom attributes being sent through when a User object is built.

When considering testing in production and the need to constantly run regression tests in production, another use case for segments would be to separate your internal traffic from your real customers. One way this can be done is by defining a segment based on the user's IP address. Any user with the company's IP address would be put within the segment as internal traffic, and an opposite rule could be implemented for the traffic coming from outside of your company. Using a segment in this manner can allow all experiments to be configured, with the first targeting rule to be that the only the external traffic segment is used to serve variations to customers.

Next, we will look at how to manage both **users** and **segments** within LaunchDarkly. The application offers several features for managing users and provides useful insight into the data being used for targeting, as well as the outcome of the feature flag evaluations.

How to manage users and segments

To manage users within LaunchDarkly, there is a dashboard that offers similar management functionality to that of feature flags, as seen in *Chapter 9, Feature Flag Management in Depth*. For the screenshots in this section, the code samples we used in the previous section to show how to build user objects have been used, with an additional user taking my name. These users can be seen here:

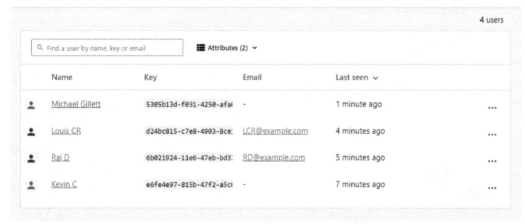

Figure 10.1 – The Users dashboard

By default, the **Users** dashboard shows all the users that have experienced a feature flag evaluation within the last 30 days. It presents user information such as **Key**, **Email** address, and when the user was last seen within any application calling the environment within the selected project. In this case, I am still using the default project and test environment.

As can be seen from this view, neither **Kevin C** nor I have email address attributes. That is because for my user, I did not provide one, but for **Kevin C**, I had set the email address to be private. Targeting using the email address would not be possible against my user but would be possible for Kevin. This might not be apparent, but we will soon see how this difference can be seen within LaunchDarkly.

The default set of attributes might not be that useful within the context of your applications or business, especially if you are sending important custom attributes to LaunchDarkly. LaunchDarkly allows us to show all the attributes within the dashboard, including both the standard and custom ones:

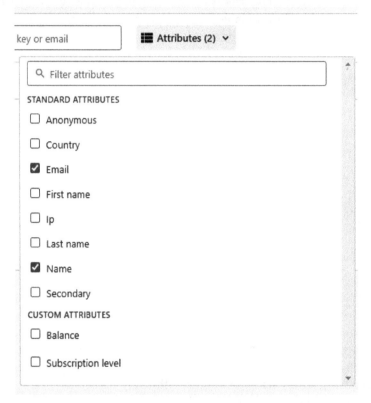

Figure 10.2 – Selecting user attributes

By accessing the dropdown menu of **Attributes**, it is possible to customize the dashboard. LaunchDarkly remembers your selection the next time you access the **Users** dashboard. In our code samples, both `Balance` and `Subscription level` were provided as custom metrics, but only for a single account. Hence, LaunchDarkly will show nothing when a user does not have an attribute set, as shown for **Louis CR** and **Raj D**:

	Name	Key	Email	Subscription level	Last seen ⌄
👤	Louis CR	d24bc815-c7e8-4903-8c	LCR@example.com	-	45 minutes ago
👤	Raj D	6b021924-11e6-47eb-bd	RD@example.com	Premium	45 minutes ago

Figure 10.3 – Showing optional custom attributes

The preceding screenshot shows how **Raj D** has a **Subscription level** while **Louis CR** does not have one set. If needed, targeting can be done against a non-set or null value for an attribute, allowing both positive and negative rules to be used.

The **Users** dashboard can be used for two functions: viewing and deleting users as required. We've already seen the functionality that's available for viewing the list of users, but it is also possible to search for specific users. Before we look at the information LaunchDarkly provides when looking at a single user, I want to discuss how to delete a user. This can be achieved by clicking the 3-dot (**…**) button on the right-hand side of the **Last seen** information; this menu only provides the option to delete a user.

It is not often that users need to be deleted from LaunchDarkly, but it can happen. This can occur with test accounts or stakeholders where the experience of seeing how a new LaunchDarkly user is to be reviewed. If dummy data has been used initially, then deleting the user is one way to start afresh. The reason why deleting a user is not often necessary is that individual flag evaluations can be set for specific users. This is done when viewing a single user, which we will look at next.

Viewing a single user

When viewing a user, the page is split into two sections: the attributes and the flag settings. First, we will look at **Attributes**:

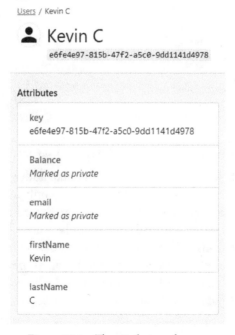

Figure 10.4 – The attributes of a user

When looking at **Kevin C**, we can see the `guid` property that was used as the **key** attribute for the `User` object and both the **firstName** and **lastName** attributes that were provided. The two-name attributes are used as the full name that can be seen at the top of the page, as explained earlier, which helps with managing users more than anything.

The other attributes that can be seen but with redacted values are **Balance** and **email**, which were both configured to be private attributes within the application. Having the attributes presented in this manner shows that they can be used in targeting but are just not visible within LaunchDarkly. This differs from an account that does not have an attribute provided, as shown in the following screenshot:

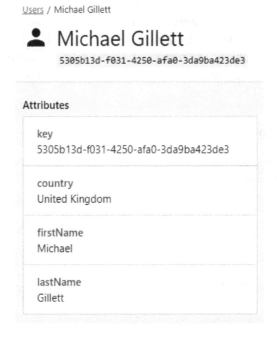

Figure 10.5 – When attributes aren't set, they do not appear

For my user, I did not create the `User` object with an email address attribute, so it does not appear at all on the user overview. Targeting for flags is only available for those attributes that are logged on a user, including those where the value is hidden, so being able to view the attributes on a user within LaunchDarkly is useful. When working with new or updated implementations, it is useful to check the data coming through for users to ensure that the planned targeting can be configured against the attributes of the user. To help with this is another LaunchDarkly feature, called **Debugger**, which will be explored in more detail in *Chapter 12, Debugger and Audit Log*.

The other part of the user page, **Flag settings**, shows the feature flags and their evaluations for a user, as shown here:

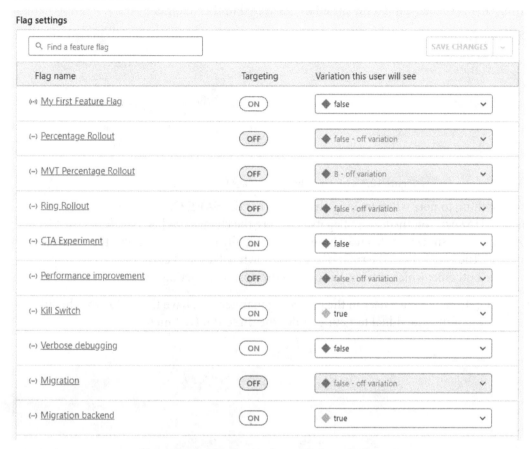

Figure 10.6 – The Flag settings page for a single user

This panel shows all the flags within the project and shows what variation is being served to the customer. All the flags can be searched, which is useful when you're looking to set a specific to a particular outcome – perhaps for a stakeholder to experience a new feature within the application. The value that's provided by a flag can only be configured when **Targeting** is enabled; otherwise, the outcome is shown as a grayed-out dropdown on the right.

Changes to the variation that are served to a user can be configured by changing the value in the dropdown menu, but they must be saved before they take effect. The UI of the flag setting changes when a different variation is selected in this manner:

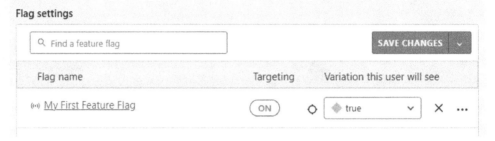

Figure 10.7 – Making changes to flags for a specific user

The first thing to note when making changes is that the **SAVE CHANGES** button is now enabled. However, on the feature flag itself, where the change has been made, a target icon now appears to show that this user is being individually targeted to achieve the selected outcome. The **X** button to the right of the dropdown will clear the user from receiving a specific outcome, which returns the user to the normal targeting rules.

In addition to removing a user in the preceding manner, you have the option to schedule the removal of the individual targeting under the three-dot (**…**) menu:

Figure 10.8 – Scheduling the removal of individual targeting for a user

The individual user screen within LaunchDarkly might not be used often, but it is useful when you're examining what attributes are being recorded and what their values are. The page's functionality to target the user with specific variations for all feature flags can be a quick way to ensure that a key user is receiving a certain experience across the product. This is great when you want to work on a per-user basis. Next, we will look at how we can group customers better based on common attribute values.

Working with segments

Often, within applications, there are common groups of customers and regularly used targeting rules. Rather than copying the rules across multiple feature flags, it is easier to label a certain group of users and use targeting rules to establish their membership. Within LaunchDarkly, these groups of users are called **segments**. We will look at managing them in this section. There is a dashboard in LaunchDarkly where we can create and view all **segments**:

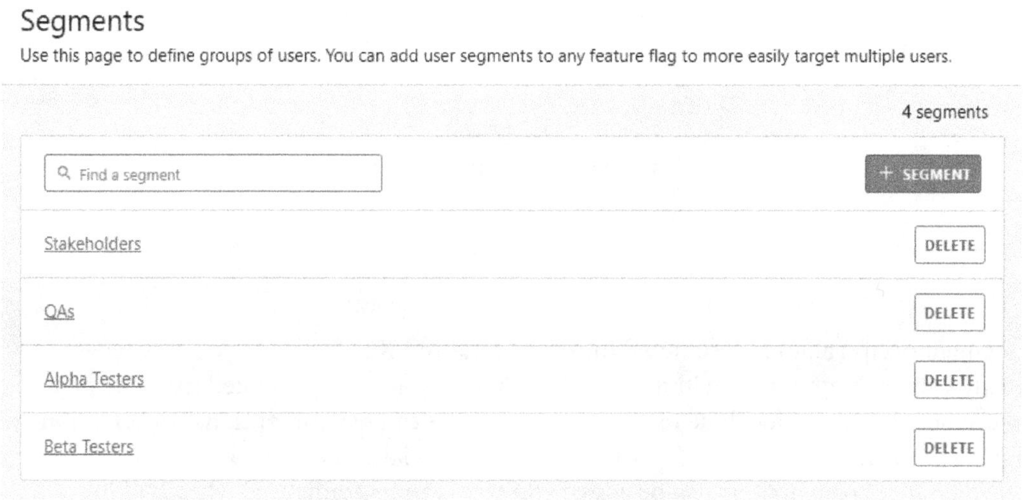

Figure 10.9 – The Segments dashboard

In the preceding screenshot, four segments exist, and in this example, they are used for testing purposes. **Stakeholders** and **QAs** are those in the company who are responsible for testing and signing off, whereas **Alpha Testers** and **Beta Testers** are customers who can provide feedback on the new features that are exposed to them.

Creating a new segment is easy as they only have a few fields:

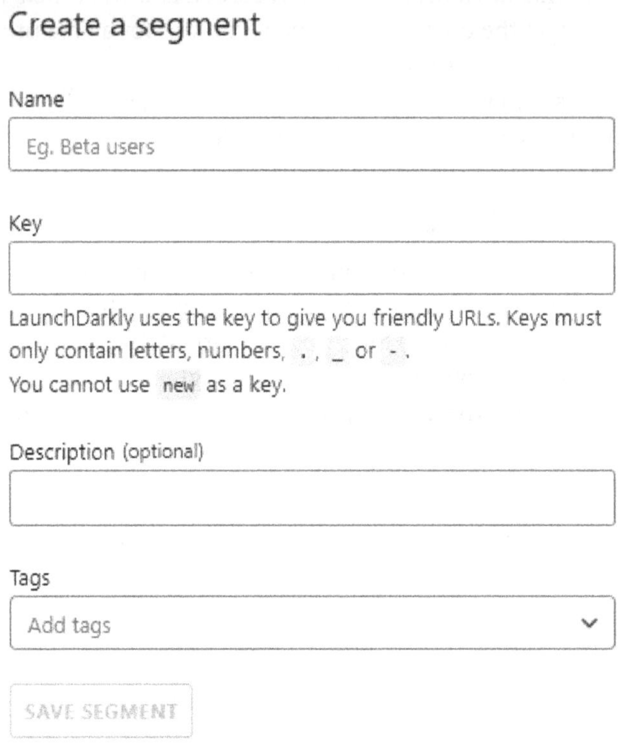

Create a segment

Name

Eg. Beta users

Key

LaunchDarkly uses the key to give you friendly URLs. Keys must only contain letters, numbers, . , _ or - . You cannot use new as a key.

Description (optional)

Tags

Add tags

SAVE SEGMENT

Figure 10.10 – Creating a new segment

A segment needs values to be entered into the **Name** and **Key** fields. As you may have already seen with other keys within LaunchDarkly, they are autogenerated from the name but can be set to any value. Like feature flags, segments can have an optional **Description** and **Tags** added, which is a great way to provide additional context for the purpose of the segment. Once created, users can be added and rules can be configured when looking at a single segment; for example, the **QAs** segment:

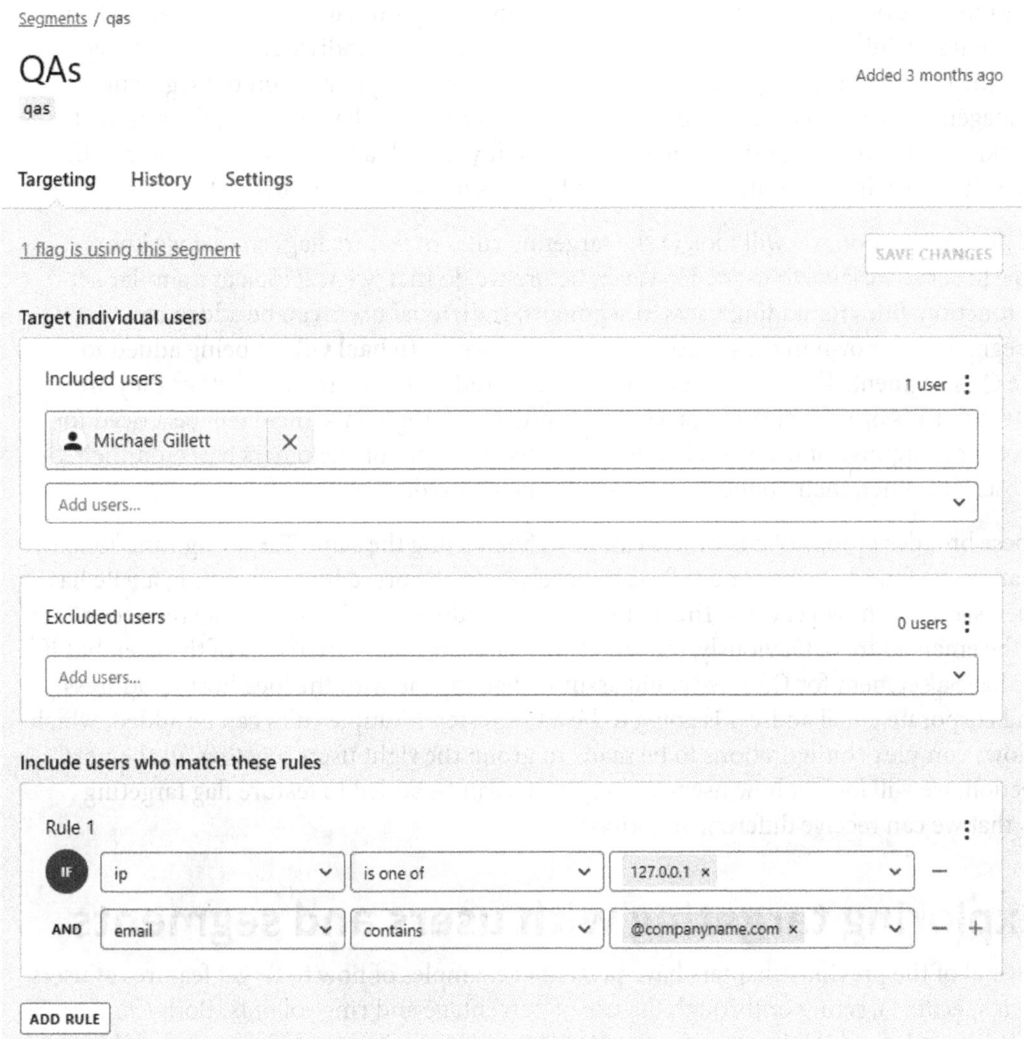

Segments / qas

QAs

Added 3 months ago

qas

Targeting   History   Settings

1 flag is using this segment

SAVE CHANGES

Target individual users

Included users   1 user ⋮

👤 Michael Gillett   ✕

Add users...   ⌄

Excluded users   0 users ⋮

Add users...   ⌄

Include users who match these rules

Rule 1   ⋮

IF   ip   ⌄   is one of   ⌄   127.0.0.1 ✕   ⌄   —

AND   email   ⌄   contains   ⌄   @companyname.com ✕   ⌄   — +

ADD RULE

Figure 10.11 – Managing a segment

The interface of the segment management screen is very similar to that of a feature flag, and that's mainly because a segment is an abstraction of the individual user targeting and the broader targeting rules functionality. The first thing to note on the segment management screen is that it is possible to view all the flags that are using the segment. Clicking **1 flag is using this segment** will present you with a list of the flags that can be navigated to help you configure the targeting rules of the flag if needed.

In the next section, we will look at the targeting rules of feature flags so that we know how to serve variants to users. However, before we do that, we will look at a similar set of functionalities for adding users to segments. Individual users can be added to a segment, as shown in the preceding screenshot with **Michael Gillett** being added to the **QAs** segment. This is how key people can be added to a segment, whether they are internal users or your customers. However, often with segments, there will be a need for broader groupings of users, such as to group certain types of customers based on their subscription tier, their country, or the device being used.

These broader groups of customers can be defined using the same **Targeting** functionality that can be found on the feature flags themselves. In the preceding screenshot, a rule has been set up with two checks. The first is for the IP address of the user, while the second is the email address. Obviously, the targeting rules can be any attributes of the user, but if we have a segment for **QAs**, we could assume that anyone with the localhost IP address and corporate email address is going to be a QA tester. Multiple rules can be added, which allows complex configurations to be made to group the right users together. In the next section, we will look at how users and segments can be added to feature flag targeting so that we can receive different variations.

Exploring targeting with users and segments

Several of the previous chapters have presented examples of how to target features at users, be it specific targeting or through the use of percentage and ring rollouts. Both *Chapter 3, Basics of LaunchDarkly and Feature Management,* and *Chapter 4, Percentage and Ring Rollouts*, provided good examples of targeting users. In this section, more details and examples will be provided to show off more use cases of LaunchDarkly's targeting.

With regards to targeting a feature flag's variation at specific individual users, there is not too much to say as there is a single configuration option. For each variation that's available on the flag, there is the option to add a specific user to receive one of the outcomes. This is true of all types of feature flags, including multi-variant ones, as seen here:

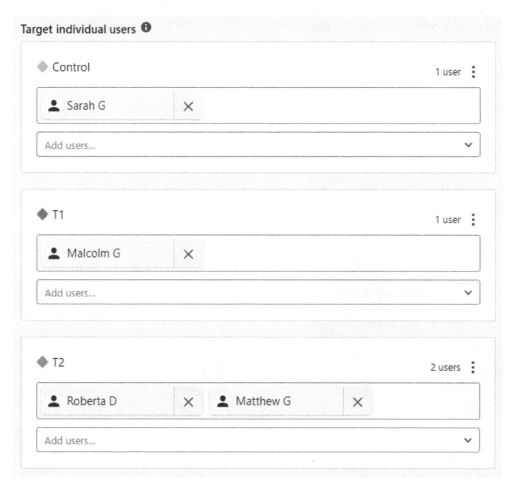

Figure 10.12 – Targeting specific users

The specific targeting is executed ahead of checking a user against any broader targeting rules. If they're found to have been set up to receive one of the variants, then the other rules will not be evaluated. The targeting rules themselves can be far more complex than targeting specific users. This can be done through multiple checks within a single rule or by having multiple rules with different outcomes. The following screenshot shows a feature flag with two rules configured:

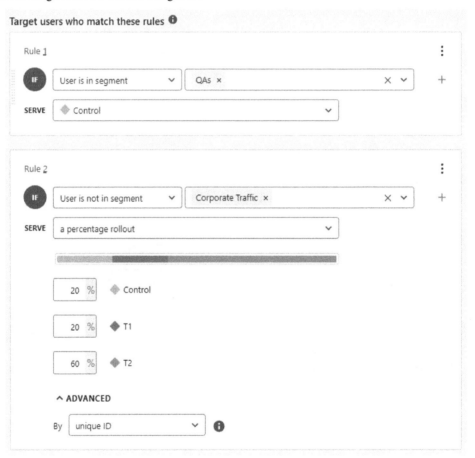

Figure 10.13 – Using multiple rules to target a feature flag

This example shows a segment or a group of users, all of whom will receive the same variation. In this example, all those users within the **QA** segment will get the **Control** variation of a feature. The second rule also relies on a segment, but when a user is not within it. In this example, the segment being used is that of detecting internal corporate traffic and targeting only those that do not fall within the segment. This approach allows you to target customers with a feature and then be able to serve a variant to those users. Here, a percentage rollout is used.

Another way of using multiple rules is to target different but similar groups of customers, such as through different outcomes per country. This can be an effective way of rolling out a feature to understand how customers within different groupings behave. Countries are a good example of where certain cultural differences or tech availability could have an impact on the success of a feature in one country versus another. LaunchDarkly allows us to have a feature turned on for some countries while off for others, and can also run a percentage rollout at the same time:

Figure 10.14 – Configuring multiple targeting rules for one attribute

By using targeting in this manner, different values of an attribute can be used for different stages of a rollout. In the preceding screenshot, the scenario could be that an experiment was already performed for the **UK** market, with **T1** being the successful variant. For customers in the **US** market, the experiment is currently being actively run and the next experiment is planned for users in **Germany**. The default targeting rule has been configured to **Control** so that customers in the other countries are getting the default implementation.

In this example, segments of customers could have been used, although the rules themselves are simple, so there might be little need to group customers in this way. However, where there are cultural or geographic needs, segments could make sense. One approach is to create segments for key business geographic regions such as America, Europe, Middle East, and Africa, Asia, and so on. Now, targeting features within a region is easier and doesn't require long lists of countries to be added to each flag. Wherever possible, it is recommended to avoid repeating work and in this regard, using segments for business-related groupings can reduce the chance of an error in the configuration and ensures that customers don't get served the wrong variant.

The targeting functionality within a feature flag is easy to work with and adding single users, groups, or new targeting rules enables opportunities for powerful targeting configurations to meet the technical and business requirements of feature management. Being able to group customers into segments can make targeting more efficient and easier to manage within LaunchDarkly.

Summary

Being able to target features at the right customers is crucial to the success of feature management and in this chapter, we covered working with **users** in greater detail than we have thus far. We began by looking at how important it is to understand how to use the User object in the code base to ensure that all the useful data can be used within LaunchDarkly's targeting functionality. Some of this data might be sensitive and LaunchDarkly caters for this data to be used for targeting, but it should not appear within the tool itself for people to view.

Then, we looked at the information that can be seen within LaunchDarkly about the user. This is useful when we want to ensure that all the correct data is coming through for users. More useful, though, is the ability to see each of the flags and which variant is being served to each user. We also looked at how it is possible to set the specific outcome of a flag for individual users.

The final part of this chapter focused on how groups of customers can be made using the targeting rules of LaunchDarkly. Using **segments** can lead to efficiencies and also reduces the risk of targeting features at the wrong customers. This chapter built on what had been explored in previous chapters but added a few new examples of how to target flags at **users** and **segments**. These examples should provide you with more ideas of how feature management can be used within your applications, as well as how the most value can be gained from this approach.

In the next chapter, we will look at the functionality provided by LaunchDarkly for working with experiments. As we mentioned previously, being able to roll out features for testing with certain groups of customers is valuable, but equally important is the ability to understand the results of the test. LaunchDarkly provides several features for getting the most out of running experiments and learning how to use this will empower your teams to optimize running tests in production.

Further reading

- The LaunchDarkly user configuration documentation details how to initialize the user object on different platforms: `https://docs.launchdarkly.com/sdk/features/user-config`.

11
Experiments

When the concept of feature management was introduced in the first section of this book, we stated that it was not a new concept. Using simple techniques such as changing a config file, we can turn a feature on or off. This is the traditional version of feature management, which is like a **switch** that enables or disables functionality for all users at once. However, with the modern functionality of feature management platforms such as LaunchDarkly, feature management can be used to concurrently serve different experiences to customers.

This ability to offer users different experiences at the same time provides us with the opportunity to experiment with customers to refine the features, journeys, and systems that are used within a product. By collecting data from the customers' usage of the product, insight can be gained into which implementations perform best for the business. This helps us find the one variant that is the most successful. This insight into customer behaviors can provide huge commercial benefits as it is difficult, or in some cases impossible, to reach these conclusions unless you perform feature management within an application.

In many ways, experimentation and the ability to test in production are the key aspects of feature management, so much so that LaunchDarkly has a section just for experiments. In this chapter, we will look at what information is provided by LaunchDarkly for managing and setting up experiments. This chapter will cover how experiments and metrics can be managed, what the different types of metrics are, and how to reach conclusions from using experiments.

In this chapter, we will cover the following topics:

- Learning about the Experiments dashboard

- Understanding metrics

- Exploring experiments

By the end of this chapter, you will have learned how to run experiments and work with several types of metrics. You should also be able to use LaunchDarkly confidently to prove when an experiment has been successful or not.

> **Note**
> Throughout this chapter, there will be screenshots and walkthroughs to show you how to use the full LaunchDarkly product. Some of the functionality that will be covered has already been detailed in other chapters, especially in *Chapter 5, Experimentation*. References will be provided for where more detailed examples or code samples have been used previously within this book. It is worth pointing out that experimentation is only available on enterprise-level plans in LaunchDarkly.

Learning about the Experiments dashboard

To begin with, we will look at the **Experiments** dashboard. This provides an overview of all the feature flags that have experiments set up for the project and the environment being used. This is also where metrics can be managed:

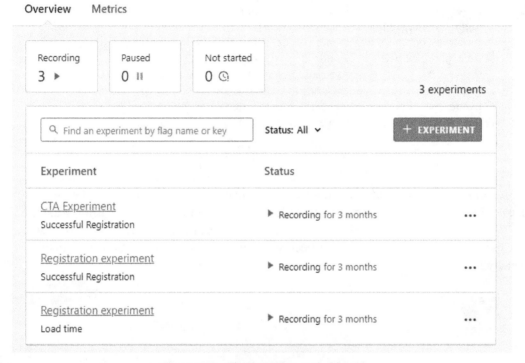

Figure 11.1 – The Experiments dashboard

The **Experiments** dashboard is comprised of two parts: **Overview** and **Metrics**. The **Overview** screen shows all the flags with experiments and provides three quick filter buttons to refine the list on the dashboard to make management easier. The three filters correspond to the states that an experiment can be in:

- **Recording**: This state means that the experiment is active and that data is being recorded. The data is being captured via the metric that was set up for the experiments.

- **Paused**: This state means that at some point, the experiment was running with data being captured, but the data collection has now been stopped.

- **Not started**: This state is for when an experiment has been set up but has never been configured to collect data against the metric.

> **A percentage rollout of a feature flag and running an experiment are not the same thing**
>
> It is worth noting that there is a difference between how a feature flag might be set up and when an experiment is being run. By this, I mean that the flag might have targeting rules configured to serve a 50:50 split between two variations, but this does not mean an experiment is running in LaunchDarkly. An experiment needs to be set to **Recording** before it is actually running. In this way, it is possible to serve multiple variants to a customer and collect data via a tool other than LaunchDarkly to gauge the success of the experiment. LaunchDarkly provides the experiment's functionality as an optional piece of functionality, which is why an experiment can be separately configured and enabled for a flag.

By default, the dashboard will show all experiments, but when one of the three buttons at the top of the screen is clicked, the filtering functionality is triggered to refine the list of flags. Clicking the **Status: Recording** filter, as shown in the following screenshot, will clear the dashboard and revert it to the default state of no status being selected. When the filter is active, it will show which status has been selected; for example, for those experiments that are recording:

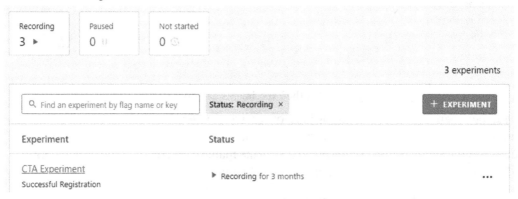

Figure 11.2 – Filtering experiments

The other way in which experiments can be refined is by searching for the name or key of the flag where an experiment has been set up. From this list, it is possible to view the status of the experiment and access some quick links from the three-dot (**...**) menu:

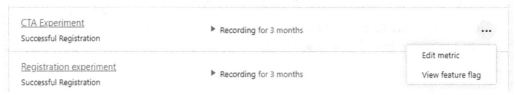

Figure 11.3 – The three-dot menu of a feature flag experiment

The menu shown in the preceding screenshot offers you a quick way to edit the metric being used to determine the success or failure of the experiment. The **view feature flag** link goes to the **Experiments** screen of the feature flag that it has been configured on. In terms of managing experiments, there is nothing else to show, and this is down to the fact that each feature flag could have an experiment set upon them, so this dashboard just collates all the flags where this is set up. The actual information of an experiment and its results can be found by viewing a feature flag.

The only other piece of functionality available from this screen is to create a new experiment. Later in this chapter, we will provide an overview of creating an experiment within a feature flag. The only difference in creating an experiment from the dashboard is that the feature flag needs to be selected where the experiment will be created. The other parts of creating an experiment will be covered in the *Exploring experiments* section.

The key to an experiment is the metric that is being used to determine the result. We will look at this next before we look at what information the experiment feature provides. While metrics were covered in *Chapter 5*, *Experimentation*, the following section will go into more detail about the variety of metrics and events that are available.

Understanding metrics

In general, when running any experiment, it is important to know what it is that is being measured to determine the success or failure of the test. There could be more than one metric being collected and analyzed, but for most experiments, there is just one. LaunchDarkly's implementation of experiments requires such a metric to determine the outcome of the investigation. The second screen of the **Experiments** dashboard shows all the metrics that have been set up for a project:

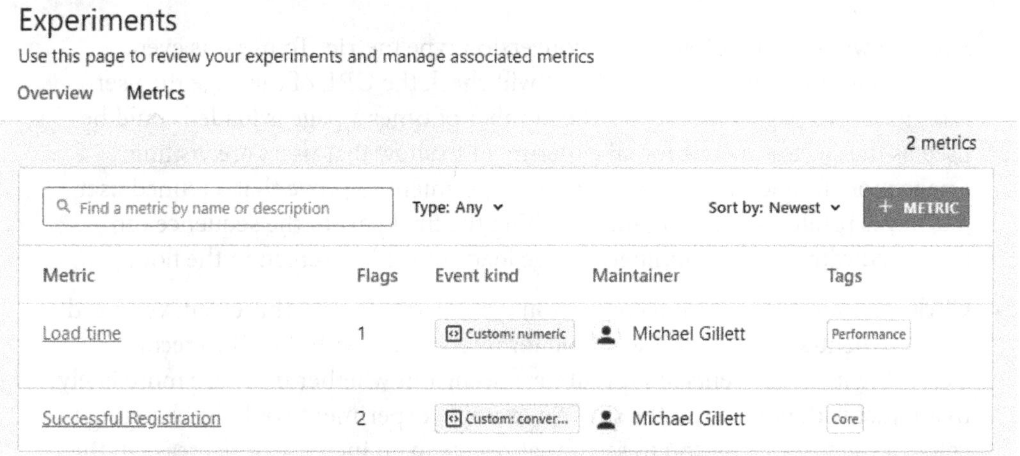
Experiments
Use this page to review your experiments and manage associated metrics
Overview
Metrics
2 metrics
Find a metric by name or description
Type: Any
Sort by: Newest
+ METRIC
Metric
Flags
Event kind
Maintainer
Tags
Load time
1
Custom: numeric
Michael Gillett
Performance
Successful Registration
2
Custom: conver...
Michael Gillett
Core

Figure 11.4 – The Metrics dashboard

Similar to the dashboards we have seen throughout this book, such as **Feature flags** and **Users,** there is the ability to search for metrics. This can be done by using their **name** or **description**. It is possible to filter the metrics and to sort the list by several attributes. The metrics themselves show how many flags have experiments set up (**Flags**) that use the metric, the type of event that the metric collects data on (**Event kind**), who maintains the metric (**Maintainer**), and the option tags (**Tags**).

Tags are useful for categorizing metrics and but cannot be used when filtering the metrics on the dashboard. For example, in the preceding screenshot, the two metrics are labeled as **Performance** and **Core**, respectively. This type of labeling can show the type of key event being tracked, with the **Load time** metric being related to the performance of the system, whereas the **Successful Registration** metric is a core metric of the business. Understanding the type of metric is important as some metrics might be temporary to determine the success of a single new feature, but others are long-lived and could have several features constantly being analyzed against a core metric.

The most important aspect of working with metrics is to understand the metric and event types that will be monitored. First, there are two types of metrics:

- **Conversion**: A conversion metric is one that registers an event for an action a user takes when a feature flag is evaluated. Examples of this are provided in the list that follows, which states the different types of events.

- **Numeric**: A numeric metric is just a number. This numeric value is important for determining the success of an experiment but is not driven by an event. Again, examples are provided in the following list of events.

These two types of metrics are then used within the following kind of events, all of which can be tracked:

- **Page View**: This kind of event is a conversion type metric. To use this event, you need to use a client-side SDK as it will check the URL of the page the user is on to record page load events. The number of times a page is loaded could be used as the success metric for an experiment to show that users are visiting a page more frequently. For example, an experiment to prove that a refined user journey is resulting in a customer reaching the final part of the sequence can be proved with a higher number of page loads of the last screen in the flow.

- **Click**: This kind of event is a conversion type metric. To use this event, you need to use a client-side SDK as it will look for interactions with the UI to record a click event. Click events are useful for finding out whether users are more likely to interact with buttons on the site. An example experiment could be changing the color or the text of a button to see if that results in an increase or decrease in the number of times customers interact with it.

- **Custom**: A custom kind of event can be either a conversion or numeric type metric. A custom kind of event does not require the client-side SDK and is used to record both page load and click events when you're not using the JavaScript and React SDKs.

  The custom conversion event can record any type of event that occurs, whether it be a page load or click – even events within systems. For example, a user might click the register button, which can be recorded, but it would be even better if we could record a custom metric on the server once a successful response has been returned for the registration attempt. While recording UI-related events is good, since it shows the intent of the customer, it can also be important to record the final event in the sequence. This is especially important when it comes to core metrics, such as registration numbers. This is because simply knowing a user's intent is not good enough as the real measure of success is whether the registration is successful or not.

  A custom numeric event can be used to record a number. With this kind of event being tracked in LaunchDarkly, it is not the event itself that is important to the success of an experiment but the value that is recorded. For example, an experiment could be run, with its success being determined by the increased number of items a customer added to their shopping cart before checking out. In this scenario, when the checkout function is called, a custom numeric event will be fired to record the number of items in the cart.

With all types of metrics and events, it is important to consider what the baseline is going to be and what an improved metric would be. When considering the conversion type of metric, LaunchDarkly will assume that a high rate of conversion is the expected outcome of an experiment. This allows LaunchDarkly to determine which variant is the successful one by identifying the one that results in more conversions per user.

However, with the numeric kind of metric, the improved metric value may be lower than the previously recorded value. This would show that a new implementation has improved the metric against the baseline. Therefore, when creating a custom numeric metric, it is possible to select the success criteria as being an increase or decrease for the recorded value.

Creating a metric

Here, we will look at creating a metric with these different kinds of events. Not many fields are required for a metric, with **Name** and **Maintainer** being the first two. It is useful to be specific when naming metrics so that it remains clear what the purpose of the metric is. Otherwise, similarly named metrics could be created that lead to the wrong ones being used over time. **Maintainer** defaults to the user creating the metric and can easily be changed if needed. The other mandatory fields relate to the event information itself; we will start by looking at the page view event kind:

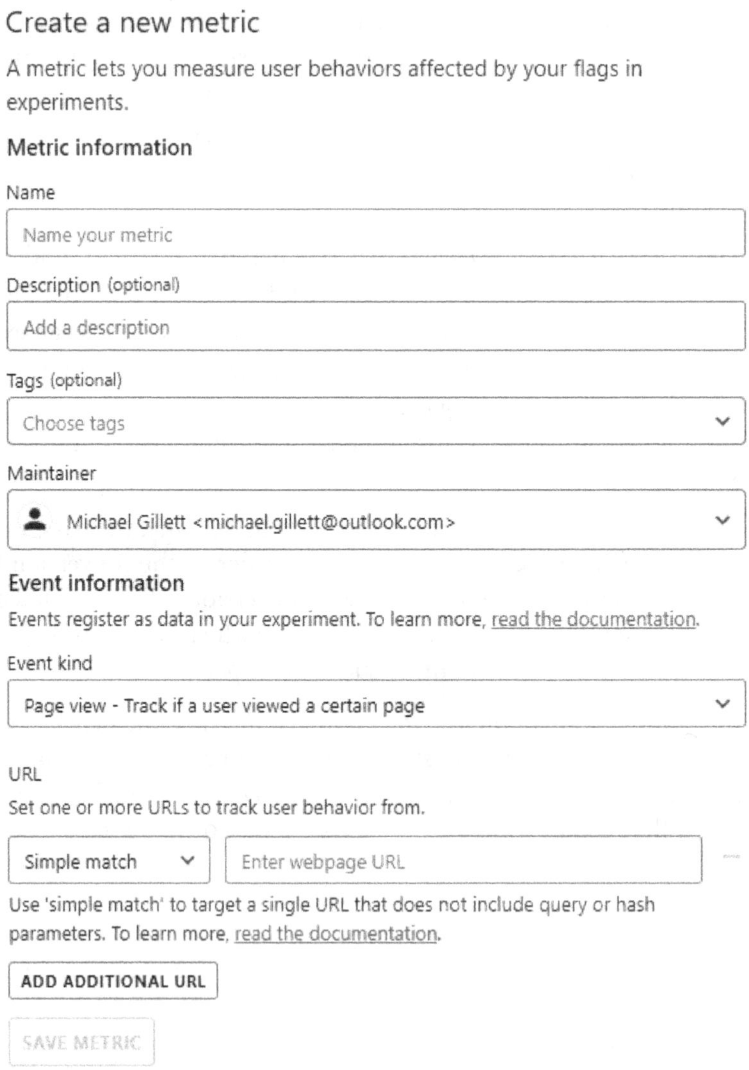

Create a new metric

A metric lets you measure user behaviors affected by your flags in experiments.

Metric information

Name

| Name your metric |

Description (optional)

| Add a description |

Tags (optional)

| Choose tags   ⌄ |

Maintainer

| 👤 Michael Gillett <michael.gillett@outlook.com>   ⌄ |

Event information

Events register as data in your experiment. To learn more, read the documentation.

Event kind

| Page view - Track if a user viewed a certain page   ⌄ |

URL

Set one or more URLs to track user behavior from.

| Simple match ⌄ | Enter webpage URL | ─ |

Use 'simple match' to target a single URL that does not include query or hash parameters. To learn more, read the documentation.

| ADD ADDITIONAL URL |

| SAVE METRIC |

Figure 11.5 – Creating a metric with a page view event

As stated previously, it is only possible to capture page view types via LaunchDarkly's client-side SDK. The SDK can fire the page load event when the URL the user is on matches the value that's been configured within the metric. There are four types of targeting mechanisms to ensure there is flexibility in how the correct URL is matched. These matching types are as follows:

- **Simple match**: URLs provided for this type are treated so that the URL itself and any parameters (hash or query string) present in the URL will all trigger the event when a user visits the URL. For example, if `example.com` was set, then the page loads event would be recorded for both `example.com` and `example.com#register`.

- **Exact match**: This matching differs from the simple match. In this case, it will not match if any additional parameters are present in the URL when a user loads the page. In this scenario, when `example.com` is set, only page loads on `example.com` would fire the event; visits to `example.com#register` would not.

- **Substring match**: This matching is used when there is a part of a route that could exist in several URLs, but they should all fire the same event. For example, if `register` were set, then `example.com#register` and `example.com/info#register` would both be URLs that the page load event would fire on.

- **Regular expression**: For more complex matching rules, it is possible to use a regular expression rule to determine a matching URL.

There are several ways of routing applications with URLs, and these four matching rules can be used for simple scenarios through to more complex ones to ensure that however the URL is being used, an event can be fired. It is possible to set more than one URL matching rule for each event to ensure that all page load triggers for an event can be used. Each URL rule that's defined on an event can use a different matching rule if needed.

The **click** event setup is similar to the **page load** setup but requires extra configuration to know what element is being clicked by the user. Setting up this kind of event looks like this:

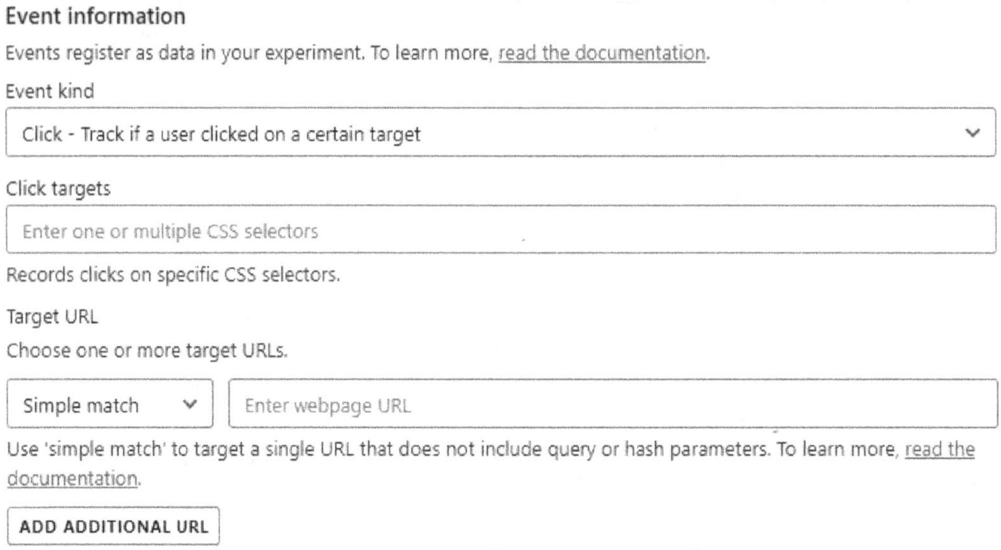

Event information

Events register as data in your experiment. To learn more, read the documentation.

Event kind

| Click - Track if a user clicked on a certain target | ⌄ |

Click targets

| Enter one or multiple CSS selectors |

Records clicks on specific CSS selectors.

Target URL

Choose one or more target URLs.

| Simple match ⌄ | Enter webpage URL |

Use 'simple match' to target a single URL that does not include query or hash parameters. To learn more, read the documentation.

ADD ADDITIONAL URL

Figure 11.6 – Creating a click event metric

The click event is triggered when a user interacts with an HTML element, usually a button or link in this scenario. To identify the element being interacted with, CSS selectors can be used. This follows the same syntax as CSS rules, so IDs or class name selectors can be used via the # or . notation; for example, `#example-element` or `.example-element`.

If a button should only be tracked on a certain page, it is possible to only have the event be triggered when the user is on specific URLs. Setting the URL matching rule can be done in the same way as with the previous page load kind of event.

When a client-side SDK is not being used or when applications without a frontend need to record metrics, then custom events need to be used. By default, the custom metric is set to a conversion one:

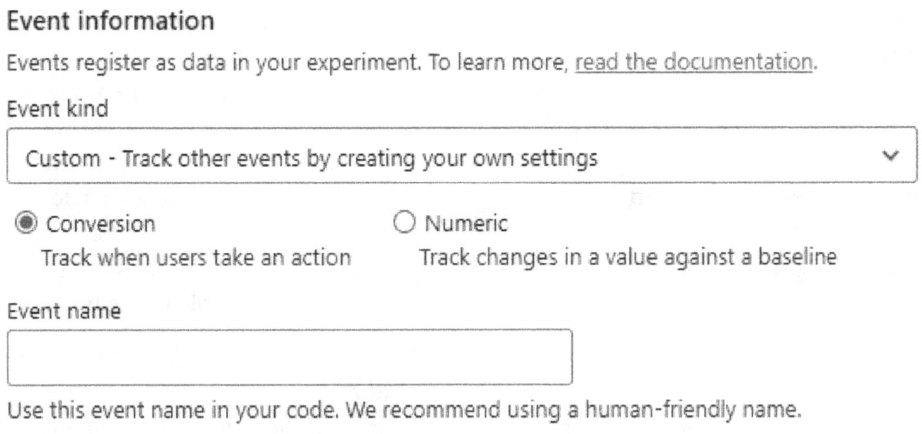

Event information

Events register as data in your experiment. To learn more, read the documentation.

Event kind

Custom - Track other events by creating your own settings	⌄

○ ⦿ Conversion
    Track when users take an action

○ Numeric
    Track changes in a value against a baseline

Event name

[   ]

Use this event name in your code. We recommend using a human-friendly name.

Figure 11.7 – Creating a custom conversion event metric

The only thing that needs to be provided for this type of event is a name (**Event name**). This name will be used within the code of the application when this event occurs. Given this is a conversion event, LaunchDarkly will count the number of times each specific event is triggered for a user per the features flags that are using this metric.

If you need to use a numeric event type of custom event, then LaunchDarkly provides more configuration options:

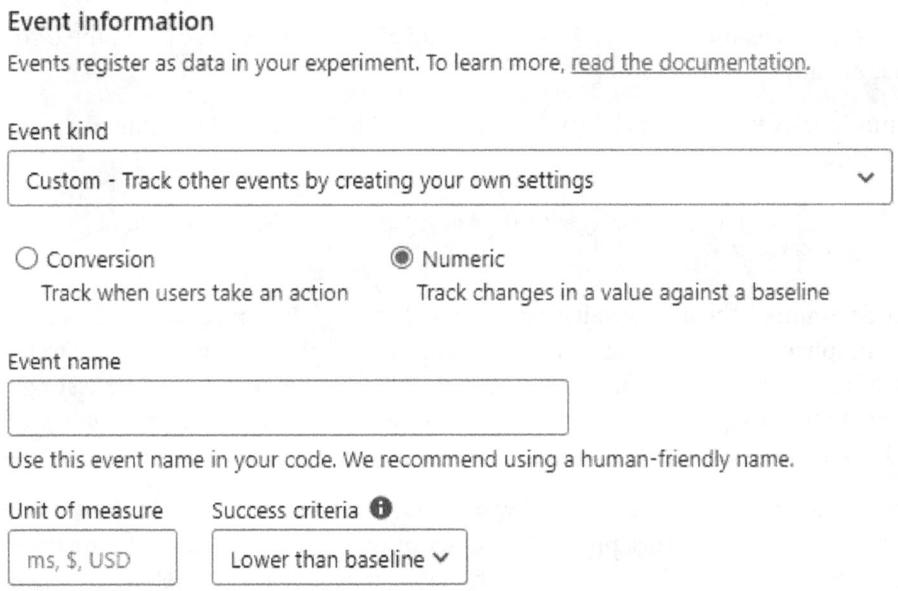

Event information

Events register as data in your experiment. To learn more, read the documentation.

Event kind

Custom - Track other events by creating your own settings	⌄

○ Conversion
    Track when users take an action

⦿ Numeric
    Track changes in a value against a baseline

Event name

[   ]

Use this event name in your code. We recommend using a human-friendly name.

Unit of measure   Success criteria ⓘ

| ms, $, USD | | Lower than baseline ⌄ |

Figure 11.8 – Creating a custom numeric event metric

For numeric types of events, there are two options. One is to provide a **Unit of measure**, which improves the experience of reviewing the data that's collected, while the other is the all important **Success criteria**. Earlier in this chapter, we looked at how the value of a numeric metric could either increase or decrease to indicate that an experiment was successful. This success criteria option is where that is set. The two options are **Lower than baseline** or **Higher than baseline**. For example, a performance-related event metric would require that the option be set to **Lower than baseline** as increased response times would be seen as successful. On the other hand, having more items in a cart before visiting the checkout would be **Higher than baseline** as this would be more valuable to the business.

In this context, the baseline is whichever variant will be served by the feature flag that should be treated as the baseline or control. This should be the existing functionality that is available to the customer. In some cases, it might mean no functionality at all, and that the experiment is to see if adding new functionality improves certain metrics. When dealing with Boolean flags, the baseline will likely be the variant that's served in the `false` evaluation. For multi-variant flags, the default one is likely to be the baseline.

With both the custom conversion and numeric events, there is a need to call the LaunchDarkly SDK's `Track()` function at the point at which the event happens. The conversion event only needs the `User` object for the person interacting with the application in that session, as well as the name of the event to be passed through:

```
_ldClient.Track("successful registration", user);
```

In this code example, the event name is `successful registration` and the `User` object has already been initialized.

For the numeric event to be tracked, the same method is called, but the value of the metric is passed through:

```
_ldClient.Track("Example event name", user, LdValue.Null,
metricValue);
```

In this code example, the metric value being passed through is named `metricValue`. With the simplicity of tracking custom events within an application, it is possible to use any type of metric as a success metric for running experiments, which enables testing to be carried out throughout the entire system and not just something that's done in the frontend and relating to the UI.

Looking back at *Figure 11.5*, there are a few optional fields, including the **Description** and **Tags** fields. As with feature flags, providing a description is useful, especially for metrics that will be long-lived. As we explored briefly in this chapter, using tags can help distinguish the use of a metric. In addition to the example of **Performance** and **Core** provided earlier, there could be tags that are used to indicate long- and short-lived metrics to help manage metrics.

From the **Metrics** section in the **Experiments** dashboard, existing metrics can be edited when needed to change most of their properties. The part that cannot be changed is the **Event kind** property of the metric.

Once the metrics have been configured, it is possible to add them to a feature flag. In doing so, this will create an experiment, which will be covered next.

Exploring experiments

While experiments have their own section in LaunchDarkly, they are not technically separate aspects of the tool. In reality, an experiment can only exist as part of a feature flag when a metric has been linked to the flag. The **Experiments** dashboard is useful for viewing all the flags that have metrics associated with them, but the real management, configuration, and overview of an experiment are found within the experimentation section of a flag. This section will recap on the information from *Chapter 5*, *Experimentation*, but more information will be provided to help you understand the results of an experiment.

When viewing the experiments section of a feature flag without any metrics attached to the flag, the only option is to create an experiment. Creating an experiment will present you with the following screen:

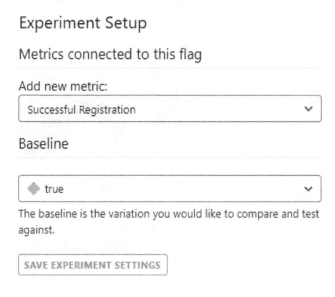

Figure 11.9 – Creating an experiment

First, a metric needs to be added that will be used to determine which variant of the flag will be the most successful. Next, you need to set the baseline variant. As explained earlier, this is often the **false** variant. It is possible to add more than one metric to an experiment that appears after saving the experiment:

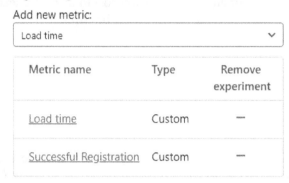

Figure 11.10 – Creating an experiment with multiple metrics attached to it

Once added to a flag, experiments will appear within their section of a feature flag, but they will not be doing anything yet. Each experiment needs to be set to **RECORDING** before data can be collected and analyzed. Once you've done this and events start being triggered, LaunchDarkly will present information about the metric:

Figure 11.11 – Viewing the data that has been collected by an experiment

In the previous screenshot, the **Successful Registration** metric is shown, where the **false** variant of the flag is being used as the baseline. The table shows a range of information about the results of how this metric performs for both variants. Let's look at this information in more detail:

- **Conversions / unique visitors**: This shows the number of times that a user triggered this event per the number of times that this variant was served. For example, the **true** row, which shows **7 / 13**, means that a total of 13 users were shown the true variant, but only 7 triggered the successful registration event.

- **Conversation rate**: The rate is then a percentage of the conversion/unique number. Using the same example of **7 / 13**, the number of people that triggered this event equates to a 53.8 conversion rate.

- **Confidence interval**: The confidence interval is a representation of what the conversion rate is really like. A broad range means that there is a small degree of confidence, while a narrow range indicates a good level of confidence in the conversion rate for the variant. This can be calculated from the information that was gathered for both variants and their outcome. In the preceding screenshot, the range of **29.1% to 76.8%** is broad, so more data should be gathered before any conclusions are reached.

- **Change**: This data shows how much better or worse the variant is performing against the baseline. In this case, the **true** variant is outperforming the **false** variant by **31.6%**.

- **p-value**: This value shows the probability of a variant impacting the users that are served that variant. This is done to measure the chance that the results of the experiment are down to random chance, rather than anything to do with the variants themselves. To elaborate further, there is always a chance that a variant performs better through luck rather than because it does anything better; this is a measure of how likely that is. LaunchDarkly treats anything with a probability of this being down to a chance of less than 5% as statistically significant, and it will show a p-value once this threshold has been met. This value becomes more accurate the longer the flag is evaluated for and the more users that trigger the events. Once an experiment has been deemed statistically significant, the results can then be trusted with a degree of confidence as proof of the experiment's success or failure. In this example, there was not enough data for LaunchDarkly to reach a statistically significant point.

Of all the data within the experiment table, it is perhaps **p-value** that is the one to be considered most important. The other numbers can fluctuate over time as external events could impact the data that's been collected, but the confidence of the test should only continue to become more precise over time. This value is the one that shows us if we can trust the rest of the data that's been gathered for the experiment, and until it is less than 5%, it might not be worth considering the test's data as meaningful. As the experiment progresses, it can be interesting to see how the results look, but it can be easy to get hung up on a result before the experiment has finished running.

In the preceding screenshot, below the table of data, there is a graph that shows how the variations have been served over time. This is hidden by default. There is a button called **Expand chart** that will reveal this graph. This is useful when you want to ensure that your ring and percentage rollouts are hitting the number of customers that's expected. In the preceding screenshot, you can see that for a while, both **true** and **false** were being served until a point where only **true** was being served.

Functionality is available for managing metrics and clearing data for experiments, but this was covered in *Chapter 9, Feature Flag Management in Depth*, in the *Learning about experiments* section. While we will not go over this information again here, it is worth bearing in mind that when an experiment is added to a flag, it is done so for all the environments of that flag. Allowing data to be recorded is done per environment, as is resetting data. The other thing to consider for flags with multiple targeting rules is that some rules can be excluded from the data that is gathered within the experiment. This is especially valuable when you're targeting specific users for testing or sign-off purposes. They will interact with the feature differently from your customers, so it would not be good to have their usage distort the real data that's been collected.

Summary

Running experiments is such a key component of feature management, and LaunchDarkly provides a good level of functionality to make the most of the opportunity that testing in this way presents. By considering the release of code as an experiment with a sense of how key metric(s) should change, teams can be data-driven in their approach to refining their product.

Separating metrics from experiments and flags themselves emphasizes the importance of knowing what it is that is being measured for an experiment. Once that is known, it is easy to determine whether that metric should increase or decrease to conclude that the experiment has been a success. LaunchDarkly leads us into thinking along these lines when creating a metric, and the separation of metrics from a single flag shows how important a metric could be for multiple features.

You should now be able to create metrics and add them to feature flags to understand which variants of a feature perform the best. You should also be able to trust the experiment's data and system to know when an experiment has reached statistical significance. With this information and the information provided in *Chapter 5, Experimentation,* and *Chapter 9, Feature Flag Management in Depth,* you should be able to run experiments across all aspects of your systems, as well as make full use of LaunchDarkly to draw confident conclusions about what works best for your customers and products.

In the next chapter, we will look at the functionality LaunchDarkly provides to help us understand how flags are being evaluated and what changes are being made within the tool. Through the use of a **debugger**, insight can be gained into how things have been configured. This is especially useful when running experiments and wanting to know if targeting is working as expected. Using **audit logs**, it is possible to quickly find the changes made to a flag.

12
Debugger and Audit Log

So far, this book has explored the theoretical uses of feature management and the ways in which LaunchDarkly can be used in these scenarios. In many of these scenarios, feature management has proven to be a methodology that reduces the risk of deployments and new code releases. Therefore, with the requirement of reducing risk, confidence in the feature management tool is crucial, which is why any good feature management tool needs to be trustworthy. LaunchDarkly offers two key functionalities that provide transparency and visibility regarding what is happening within the system.

The first piece of functionality is the **Debugger**. This offers a look at what LaunchDarkly is evaluating for feature flags, users, and experiments. The second component is the **Audit log**, which provides an overview of all the changes made within LaunchDarkly. Together, they give you the insight that is needed to trust the system, as the **Debugger** allows teams to understand how the system is working and if things have been configured correctly, and the **Audit log** gives reassurance that configuration changes are correctly logged.

In this chapter, we will explore both the **Debugger** and **Audit log** sections and understand what information they impart. While the functionality provided by both sections of the site does not directly form part of feature management, they are important to consider when using LaunchDarkly. This is especially true with more complex and more important applications where better tooling and governance become increasingly important. Without the **Debugger**, it can be difficult for you and your teams to understand how LaunchDarkly evaluates flags and what kind of user and experiment data is sent to the tool. Without the **Audit log**, there could be a risk that there was no record of the changes being made to the targeting and configuration of the flags. This would be unacceptable if LaunchDarkly was to be used in large and important scenarios where an insight into the changes being made is crucial.

In this chapter, we will focus on the following topics:

- Understanding the **Debugger** and **flag events**
- Exploring users within the **Debugger**
- Viewing experiments within the **Debugger**
- Understanding the **Audit log**

In this chapter, there will be screenshots of LaunchDarkly but no code samples, as all the functionalities being described are contained within LaunchDarkly itself.

Understanding the Debugger and flag events

The **Debugger** can be accessed via the LaunchDarkly navigation bar. When you first open it, it will only show a little information because it is a live stream of data that any LaunchDarkly clients using the SDK key will connect to, and so the data will be streamed to this page over time. There are three sections to the **Debugger**: **Flag events**, **User events**, and **Experimentation events**. To begin with, we will examine the **Flag events**:

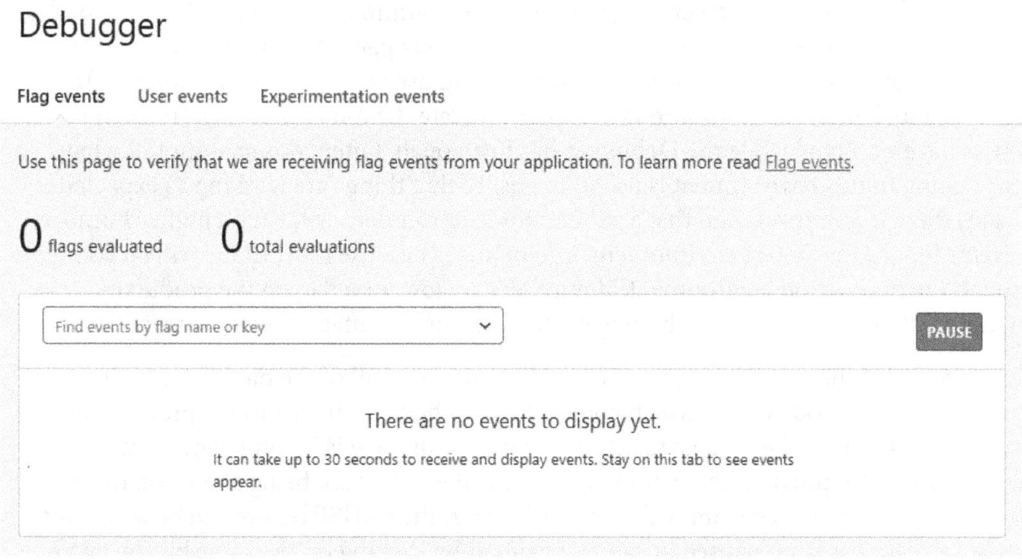

Debugger

Flag events   User events   Experimentation events

Use this page to verify that we are receiving flag events from your application. To learn more read Flag events.

0 flags evaluated   0 total evaluations

| Find events by flag name or key ⌄ | PAUSE |

There are no events to display yet.

It can take up to 30 seconds to receive and display events. Stay on this tab to see events appear.

Figure 12.1 – The Debugger

As shown in the preceding screenshot, the **Debugger** section can be used to check which flag evaluations are occurring. This is especially useful for teams. And, as we will learn in this section, there are a few key pieces of information that are provided by the **Debugger**. The **Debugger** starts working once the page is opened and is a frontend view of a stream of data. Sometimes, it takes a little while for the data to start appearing on the page. Once **flag events** start coming through, it is possible to view them all on this page:

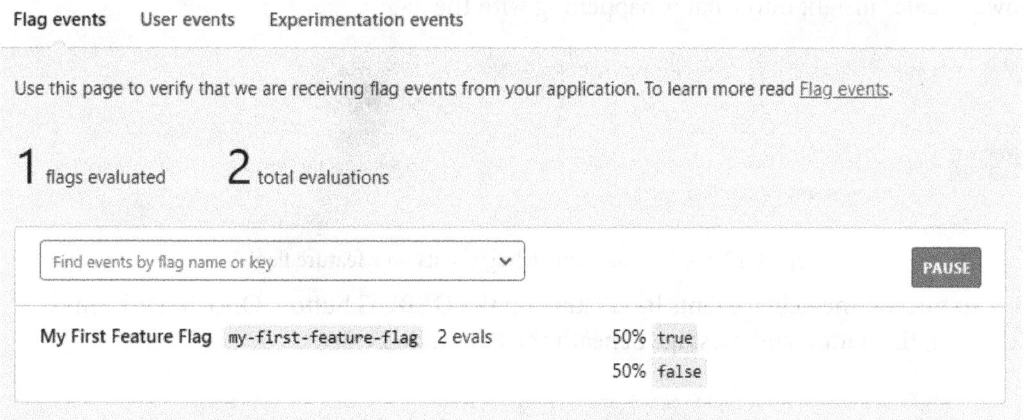

Flag events   User events   Experimentation events

Use this page to verify that we are receiving flag events from your application. To learn more read Flag events.

1 flags evaluated   2 total evaluations

| Find events by flag name or key ⌄ | PAUSE |

My First Feature Flag `my-first-feature-flag` 2 evals   50% `true`
50% `false`

Figure 12.2 – The Debugger with flag events

At the top of the preceding screenshot, you can view a summary of the total number of flag events and evaluations. These are events that have been triggered within the LaunchDarkly SDK for the project and the environment that is being used to access the **Debugger**. The **flags evaluated** and **total evaluations** numbers indicate the total number of times the metrics have occurred while the **Debugger** has been open. Often, this snapshot of what is happening in this environment is useful to ensure that things are working as expected. Usually, there is an expectation that a production environment will have a higher number of events logged than a test environment. Often, this is because most traffic will be coming through the production environment. If numbers are low for a flag on the production environment, it could immediately suggest that the implementation is not correct.

As mentioned earlier, the **Debugger** shows a live stream of all of the data coming through from the LaunchDarkly client, which could be a lot when coming from the production system to this page, so being able to select a single feature flag is helpful. However, LaunchDarkly also provides the option to pause the live data from being shown on this page. To stop showing the data stream within the **Debugger**, the **PAUSE** button can be used. Once the data stream has been paused, it can be resumed by clicking on the same button again. In the screenshots shared in this chapter, the volume of data coming through is minimal.

In *Figure 12.2*, underneath the highlighted numbers, there is a list of all the feature flag events. Each flag event shows the name, the key, the total number of evaluations during this **Debugger** session, and percentages of the variations being served. It is possible to select a single flag from the presented list to view its information, which is useful when attempting to understand how that flag is functioning.

Additionally, LaunchDarkly shows an option to turn on debug events for a single flag to show a greater insight into what is happening with the flag:

Figure 12.3 – To enable the debug events on a feature flag

You can turn on the debug events by clicking on the **DEBUG** button. Once it has been clicked on, the button and message beneath the flag will change:

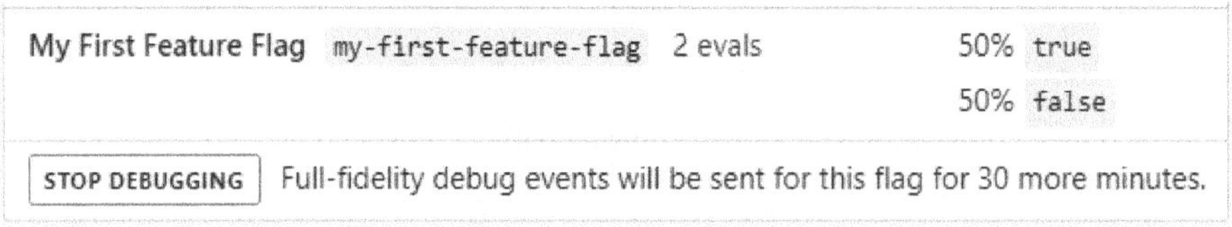

My First Feature Flag   my-first-feature-flag   2 evals   50% true
                                                                 50% false

STOP DEBUGGING | Full-fidelity debug events will be sent for this flag for 30 more minutes.

Figure 12.4 – Disabling the debug events on a feature flag

To stop debugging for a flag, you need to click on the **STOP DEBUGGING** button. This extra data is only useful for debugging, so LaunchDarkly restricts it from being sent for more than 30 minutes. This debug data will become expensive if we simply keep the mode enabled for the flag, as it will consume unnecessary bandwidth and storage. So, to avoid incurring heavy expenses, the default debug time is set to 30 minutes. If needed, once the 30 minutes have elapsed, the debugging mode can be re-enabled for a flag to continue investigating.

The debug mode shows each evaluation for each user on that flag. When clicking on a specific evaluation, it will even show any additional attributes of that user. This information is useful when you want to understand why users receive a specific variation of the flag, especially if it does not appear to be evaluating as expected. The flag evaluation events appear as follows:

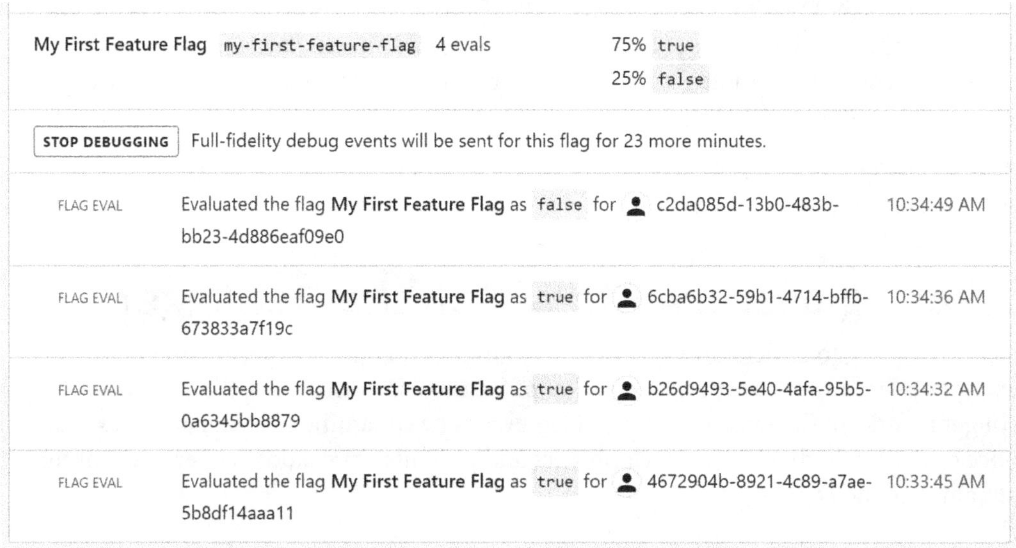

| My First Feature Flag | `my-first-feature-flag` | 4 evals | | 75% | `true` |
| | | | | 25% | `false` |

| **STOP DEBUGGING** | Full-fidelity debug events will be sent for this flag for 23 more minutes. |

FLAG EVAL	Evaluated the flag **My First Feature Flag** as `false` for 👤 c2da085d-13b0-483b-bb23-4d886eaf09e0	10:34:49 AM
FLAG EVAL	Evaluated the flag **My First Feature Flag** as `true` for 👤 6cba6b32-59b1-4714-bffb-673833a7f19c	10:34:36 AM
FLAG EVAL	Evaluated the flag **My First Feature Flag** as `true` for 👤 b26d9493-5e40-4afa-95b5-0a6345bb8879	10:34:32 AM
FLAG EVAL	Evaluated the flag **My First Feature Flag** as `true` for 👤 4672904b-8921-4c89-a7ae-5b8df14aaa11	10:33:45 AM

Figure 12.5 – Viewing the debug events of a feature flag

Being able to investigate the data that LaunchDarkly has received about the user and what variation was served to that customer is important. To trust the feature management process, it is important to ensure that the user has the necessary information on their profile, that the user is encountering the flag within the execution of the application, and that the targeting rules are working. The **Debugger** can provide this insight.

The **Debugger** does not help you to identify where an issue or misconfiguration exists, but it can help to prove that the implementation or configuration of a flag is not correct. There are many ways in which a user might not receive the expected flag evaluation, but that could be down to some common issues, such as the following:

- The implementation of the `User` object is not correct.

- Not identifying the user in a consistent manner (for example, using a hash cookie).

- Not providing the correct value for the `User` object attribute.

- Not having the feature flag evaluation executed at the expected point in the runtime of the application.

- Not having the targeting rules enabled, or configured correctly.

When things are not working as expected with a feature flag, it is usually a setup or logic issue within the code base. This is because the code for a LaunchDarkly flag is simply an `if` statement with a remote evaluation of a `true` or `false` value (or a multivariate value). In this instance, the **Debugger** is useful to help identify where the issue might be.

In the next couple of sections, we will explore what information the **Debugger** can offer to deal with **users** and **experiments**.

Exploring user events within the Debugger

When attempting to understand what LaunchDarkly is doing with specific **users**, the **Debugger** offers a similar experience to that of feature flag evaluations. Again, the **Debugger** works in the same manner as **Flag events** by streaming live data and presenting it once opened. It highlights the key numbers and presents a breakdown of each event that is streamed to the **Debugger**:

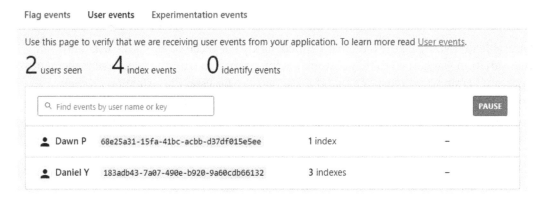

Figure 12.6 – Viewing users within the Debugger

The **Debugger** records the following three types of **user events** to demonstrate what is happening within your application and the tool itself:

- **users seen**: This shows the number of unique users that are seen during the time the **Debugger** is open.

- **index events**: This indicates the total number of index events for users. It is triggered by the LaunchDarkly SDK for feature flag evaluations.

- **identify events**: This shows the number of identity events that are triggered by explicit methods within your code base.

Similar to the **flag event** totals, these numbers can quickly show whether an application is working with LaunchDarkly correctly. For example, if the number of users seen is not showing an expected level, then it might suggest that something has not been implemented correctly or that too many users are being labeled as anonymous.

Figure 12.6 shows the aforementioned numbers plus the list of users. Clicking on a user reveals the events being captured about them via the SDK. In this example, the events for a user named **Daniel Y** were being indexed by LaunchDarkly. Clicking on a single event for that user reveals all the information about that user and the event:

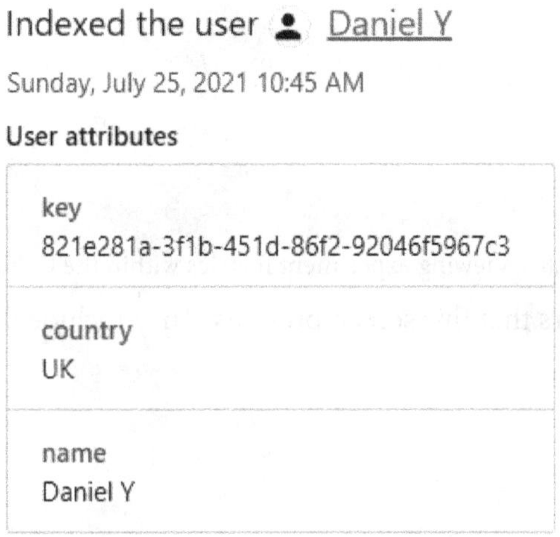

Indexed the user & Daniel Y

Sunday, July 25, 2021 10:45 AM

User attributes

key
821e281a-3f1b-451d-86f2-92046f5967c3

country
UK

name
Daniel Y

Figure 12.7 – Viewing all the data for the user event logged

The information shown in the preceding screenshot might differ from the user itself, as this is a view of data used for that particular event. There might be a need to view the single user, and this can be accessed by clicking on the username. This reveals all the information LaunchDarkly has on the user, as explored in *Chapter 10, Users and Segments*.

As with flag events being able to view the **user events** within, the **Debugger** offers visibility on how your application and LaunchDarkly are working. This view helps you to identify issues and can save time when you are investigating where those issues might be. The final part of the **Debugger** is viewing **experiment** data, which we will cover next.

Viewing experiment events within the Debugger

It is important to be able to confirm that experiment metric events are being tracked correctly by your application and that LaunchDarkly is recording them as expected. The **Debugger** shows summarized information for the metric events, just as it does with flags and users:

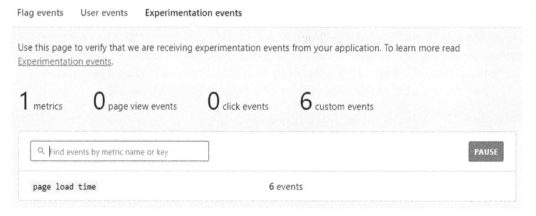

Figure 12.8 – Viewing experiment metrics within the Debugger

There are four key numbers that this screen presents. They include the following:

- **metrics**
- **page view events**
- **click events**
- **custom events**

The first is the total number of metrics detected during the session. The other three numbers relate to the types of metrics. These types of events were detailed in *Chapter 11, Experiments*. Viewing the totals of the metric types in this way, again, provides a quick overview of whether things are working as expected. If the **Debugger** is used on a test environment after implementing the tracking of a new metric, it can be quickly validated in LaunchDarkly by monitoring the events correctly.

Underneath the metric numbers is a list of the metric events that the LaunchDarkly SDK has tracked. *Figure 12.8* shows six custom events for the **page load time** event. Clicking on the name of the tracked metric reveals the period of time that metric was tracked and for which user:

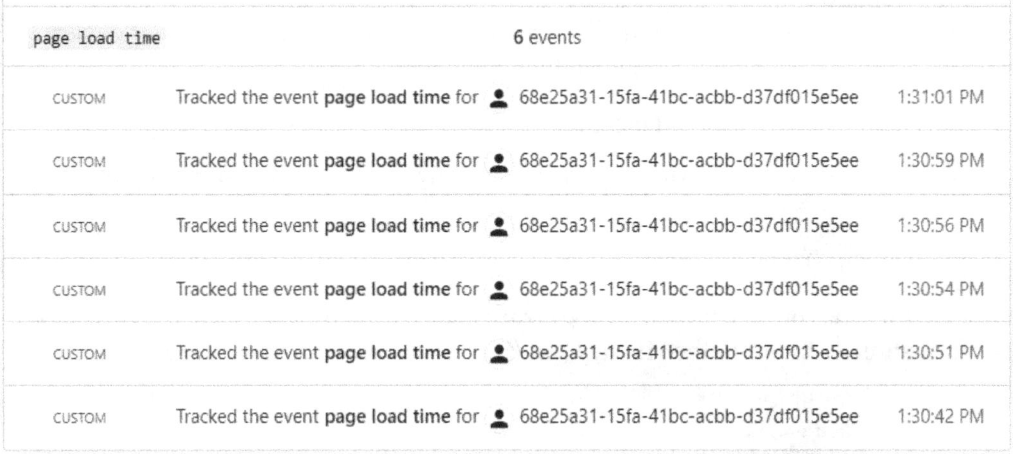

page load time	6 events	
CUSTOM	Tracked the event **page load time** for 👤 68e25a31-15fa-41bc-acbb-d37df015e5ee	1:31:01 PM
CUSTOM	Tracked the event **page load time** for 👤 68e25a31-15fa-41bc-acbb-d37df015e5ee	1:30:59 PM
CUSTOM	Tracked the event **page load time** for 👤 68e25a31-15fa-41bc-acbb-d37df015e5ee	1:30:56 PM
CUSTOM	Tracked the event **page load time** for 👤 68e25a31-15fa-41bc-acbb-d37df015e5ee	1:30:54 PM
CUSTOM	Tracked the event **page load time** for 👤 68e25a31-15fa-41bc-acbb-d37df015e5ee	1:30:51 PM
CUSTOM	Tracked the event **page load time** for 👤 68e25a31-15fa-41bc-acbb-d37df015e5ee	1:30:42 PM

Figure 12.9 – Viewing the individual events for a metric

It is useful to be able to view the individual events and see which user triggered them. Similarly to the implementation of flags and users, it could be that the implementation of a metric being tracked is not correct or that the code execution flow is not as expected. Viewing this debug information can help you to confirm whether the code is working as expected. Of course, with the event tracking implementation, it is a simple piece of code to track the metric, so it is not likely to be an issue. In the preceding screenshot, the same user was triggering this event on purpose.

The LaunchDarkly **Debugger** offers you the opportunity to view how the application and LaunchDarkly are integrated. The live stream of data through to a dashboard helps you to discover issues quickly, understand what information is missing, and what configuration has been incorrectly set up. This helps teams to ensure their implementations are correct before deploying code to production. It would not be good to discover that, once deployed to production, a rollout or experiment cannot be performed because the information is missing for the successful targeting of a flag.

Next, we will take a brief look at how LaunchDarkly provides insights into what is being changed within the tool itself.

Understanding the Audit log

In *Chapter 9, Feature Flag Management in Depth,* we looked at a historical view of all the changes that are made to an individual feature flag. We established that this was useful for audit purposes and to identify changes that were made to revert to a previous state if needed. LaunchDarkly not only provides this type of information for flags but also for the whole tool itself. The **Audit log** tracks change across a whole environment. It details information relating to feature flags, including the following:

- The creation, archiving, and deletion of flags

- Changes to the targeting rules of a flag, including the enabling or disabling of the targets

- Experiment changes to a flag, including the reset of the collected data

Additional information is recorded for the creation and deletion of segments. The view of this information is similar to the **History** section that we looked at for feature flags:

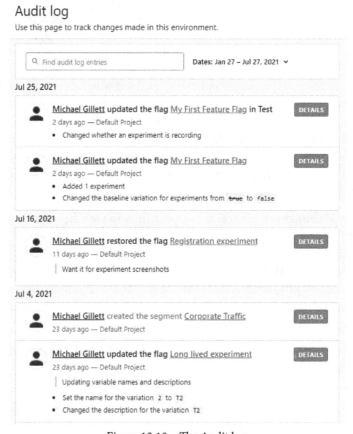

Figure 12.10 – The Audit log

The events shown in the log are sorted chronologically and are grouped by date. Given that this is a tool to view the historical changes within an environment, there are no options to view these events in any other way. There is an option to search for entries within the log via the name of the changed feature flag. There is also the ability to change the date range for the events shown in the list. Viewing an entry reveals more information about what the change entailed:

Michael Gillett updated the flag Long lived experiment

July 4, 2021 6:44 PM — Default Project

Updating variable names and descriptions

Changes:
- Set the name for the variation 2 to T2
- Changed the description for the variation T2

Patch

```
COPY
[
  {
    "op": "test",
    "path": "/variations/2/_id",
    "value": "897763cd-a3d5-4588-a7bb-574f022ade04"
  },
  {
    "op": "replace",
    "path": "/variations/2/name",
    "value": "T2"
  },
  {
    "op": "replace",
    "path": "/variations/2/description",
    "value": "The second experiment"
  }
]
```

Figure 12.11 – The details of an audit log entry

The details view shows the JSON of what was saved for the event. The ability to view this granular information can be extremely useful and, again, provides the reassurance and confidence that there is full transparency in what is going on within LaunchDarkly. In addition to the information shown in *Figure 12.11*, it is also possible to view who made the change, what flag was affected, any comments provided when the change was saved, and a readable version of the change followed by the JSON itself.

In some cases, there is a change made to the existing settings of a flag, and LaunchDarkly provides a *diff* of the JSON in this scenario. *Figure 12.11* only shows the event where a new configuration is being set, so there is no *diff*. The **Diff** tab shows the before and after of the configuration:

 Michael Gillett updated the flag <u>My First Feature Flag</u> in **Test**

July 25, 2021 10:30 AM — Default Project

Changes:

- Changed whether an experiment is recording

Patch   Diff

```
  },
  "experiments": {
    "baselineIdx": 1,
    "items": [
      {
        "environments": [],
        "environments": [
          "test"
        ],
        "metricKey": ""
      }
    ]
  },
```

Figure 12.12 – The Diff of an Audit log entry

In addition to being able to see exactly how a flag was configured and what changed over time, as is the case with the **History** section of a feature flag, the **Audit log** gives teams the confidence to know that changes are recorded and known good states can be easily returned to if needed. For some businesses, this type of extensive audit trail showing what is happening in an environment is crucial for feature management to even be considered as a way of working with production applications. LaunchDarkly's **Audit log** provides this necessary overview of the history of the changes made within an environment.

Summary

This chapter examined how LaunchDarkly can help teams understand how flags, users, and experiments are working and provide a view of all the changes. Neither the **Debugger** nor the **Audit log** is necessary for feature management, but both make the experience of working with feature management and LaunchDarkly easier.

With the **Debugger**, development teams can gain insights into how the LaunchDarkly SDK is working with the data provided. This can speed up the process of identifying implementation or configuration issues, as the **Debugger** will show you whether things were not working as expected. This is a more effective way of discovering something is wrong than just trying to discover this from the application logs themselves. With the ability to view both summary information and enable debug data, teams can gain as much information from the SDK's **Debugger** as needed.

Throughout this book, feature management has been shown to extend the practices of modern software delivery processes. In many of those processes, there is good traceability of what has happened during the build and deployment of software. Where feature management can introduce the separation of deployment from the release of software, it is still necessary to have a record of what has been changed, including when and by whom. The **Audit log** provides exactly this traceability.

In the next and closing chapter of this book, we will take a look at all the configuration options of the entire LaunchDarkly account. This will include managing teams, projects, and roles in addition to billing information, security options, and usage reports.

13
Configuration, Settings, and Miscellaneous

In this final chapter, we will explore the remaining aspects of LaunchDarkly. The functionalities that are covered here do not directly impact feature management, but they are all concerned with managing the settings and configuration of the LaunchDarkly tool.

As mentioned in earlier chapters, having good governance around LaunchDarkly is important. Feature management with LaunchDarkly offers a powerful opportunity for members of a team to make immediate changes to the production environment. However, there is a need to ensure that the right people have the right permissions to update the right feature flags.

To put the right measures and controls into place, this chapter will detail how teams can be managed and how permissions can be assigned to people in a project. This ensures that there is the confidence that only the people who belong to a project can make any changes. The segregation of flags within projects is important to help with the separation of concerns and to keep lists of feature flags maintainable. This chapter explores how you can optimize the configuration of projects and environments. The remainder of the chapter provides information on how to authorize requests to LaunchDarkly and the wider security functionality of the product. Following this, we will include information on the billing and usage reports of the tool. Understanding this is essential for keeping costs to a minimum but still gaining excellent value from the tool. The final pieces of functionality to be looked at will be your own profile within the tool and the history of all the changes within LaunchDarkly.

In this chapter, we will cover the following main topics:

- Discovering team management
- Exploring project management
- Learning about roles and permission management
- Learning about authorization and security
- Understanding billing and usage
- Exploring your profile
- Discovering history

> **Note**
>
> In this chapter, there are no code samples, as it covers functionality exclusively within the LaunchDarkly tool. There will be some overlap with concepts and examples from other chapters, but the information provided here will be more detailed than previously covered in this book.

Discovering team management

To begin, in this section, we will focus on the **Account settings** aspect of LaunchDarkly, which can be accessed via the sidebar. This page defaults to the **Team** section. By *team*, LaunchDarkly refers to all those who can log into this organization; it does not necessarily reflect all the teams that you have in your organization. In the upcoming sections, we will explore **Projects** and **Roles**, where we will take a look at how your own teams can be set up in LaunchDarkly. Underneath the **Team** section is a list of all people within the team, including a total number of members, as follows:

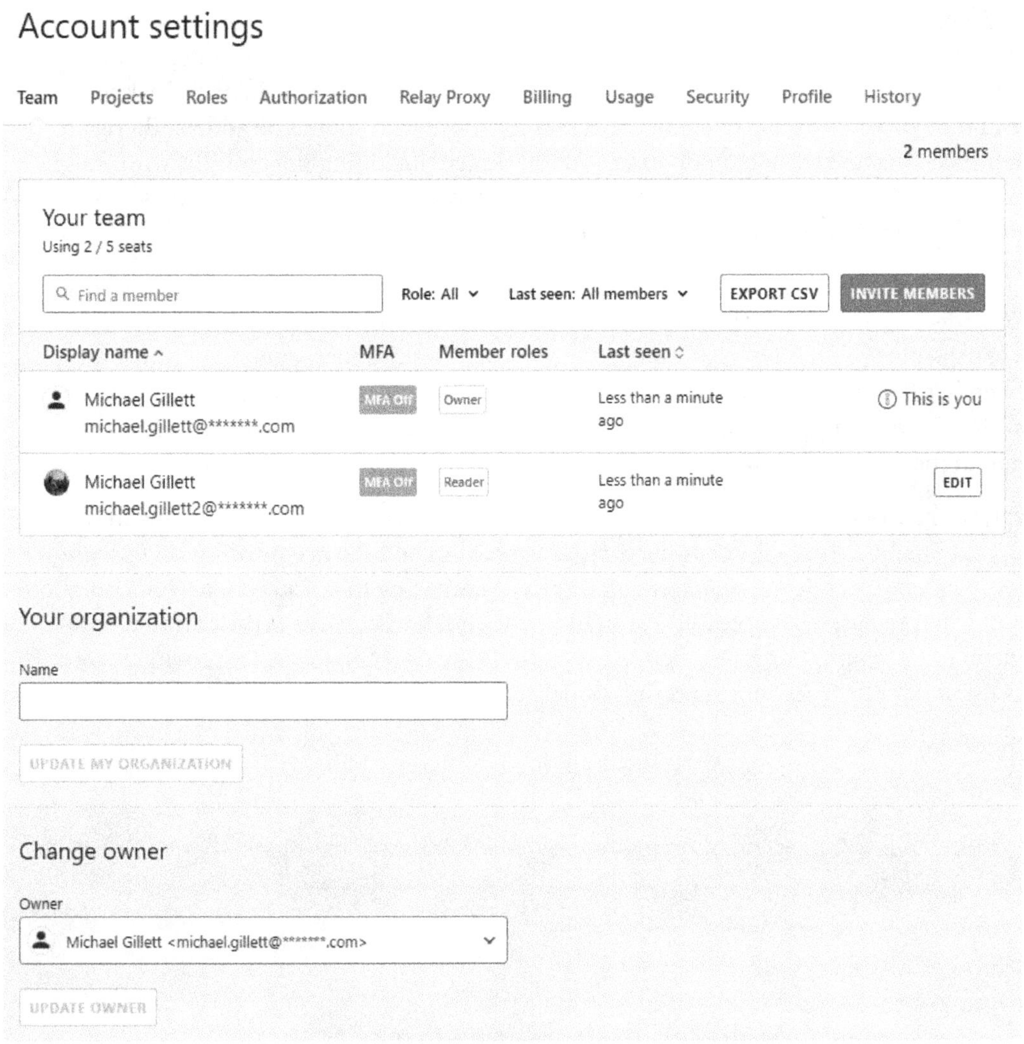

Account settings

Team   Projects   Roles   Authorization   Relay Proxy   Billing   Usage   Security   Profile   History

2 members

Your team
Using 2 / 5 seats

| Q Find a member | | Role: All ∨ | Last seen: All members ∨ | EXPORT CSV | INVITE MEMBERS |

Display name ∧	MFA	Member roles	Last seen ⟳	
👤 Michael Gillett michael.gillett@*******.com	MFA Off	Owner	Less than a minute ago	ⓘ This is you
🌐 Michael Gillett michael.gillett2@*******.com	MFA Off	Reader	Less than a minute ago	EDIT

Your organization

Name

UPDATE MY ORGANIZATION

Change owner

Owner

👤 Michael Gillett <michael.gillett@*******.com>   ∨

UPDATE OWNER

Figure 13.1 – Team management in Account settings

Looking at the preceding list of members, we find that it behaves just like the other LaunchDarkly dashboards that we viewed in the previous chapters. It is possible to search for a specific member via their name. The list can also be filtered based on the role a member has or when they last logged in using timeframes from the last 30 days up to the last year. The latter form of filtering is useful when it comes to identifying people who no longer use the tool, which is important when we come to the *Understanding billing and usage* section of this chapter.

The list of users and their settings can be exported as a CSV file. This is useful when it comes to audits if the analysis of larger lists of people is required or for integrations with other systems. It is also possible to invite other members to this LaunchDarkly account by providing a list of email addresses for those who should be added. There are other mechanisms by which individuals can be added, including integrations with **Single Sign-On** (**SSO**) and external identity providers, such as **Azure Active Directory**. These types of integrations require an enterprise plan.

In *Figure 13.1*, there are two members in this particular team, and the information that is provided is useful for managing the members and the broader account. Shown in this list are the person's name, email address, and avatar. Besides that, there is an indicator of whether the member's account is using **Multi-Factor Authentication** (**MFA**), which can significantly improve the security of access to LaunchDarkly. This is covered in more detail in the *Learning about authorization and security* section of this chapter.

Next, the role of each member is presented. The role determines the permissions that the individual has (we will cover more regarding **Roles** and **Permissions** in the *Learning about roles and permission management* section). Following this, you can view information about when the user last accessed LaunchDarkly, and, finally, there is the option to edit a member's account. Editing someone's account is the only functionality provided from this list and it offers account-level functionality:

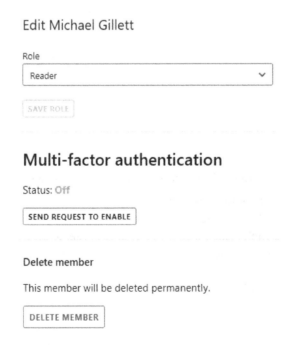

Figure 13.2 – Editing a member's account

From the **Edit** screen, it is possible to change the role that someone belongs to. If MFA is something that the teams or business requires, then the process to enable it for an account can be triggered from here. Finally, there is the option to permanently delete someone from accessing this LaunchDarkly account.

In *Figure 13.1*, underneath the list of members, there is the option to name the organization that this account belongs to. I have not set a name for the account that I have used while writing this book, but this is something that should be set for the real-world usage of LaunchDarkly. The final piece of functionality within the **Team** section is to configure who the owner is. By default, the first person to create the account is the owner, and this can only be changed by the owner account when needed.

Next, we will take a look at the **Projects** of a LaunchDarkly account and gain an understanding of the options that are available when grouping feature flags together.

Exploring project management

Projects are how feature flags, segments, metrics, and some settings can be grouped together. A **project** could be a *1:1* mapping to a team so that the feature flags for the team's products are all grouped together. Alternatively, a project could span multiple teams but remain focused on a particular value stream of the business. So in this case, we group together flags for products that form key user experiences.

In addition to the projects, there are **environments**. The environment for which a feature flag is being configured exists only within one project, and projects can have several environments set up. Both projects and environments were explained in *Chapter 3, Basics of LaunchDarkly and Feature Management*. However. in this section, I will share more information.

The following screenshot displays all projects with their environments listed within them underneath the **Projects** section:

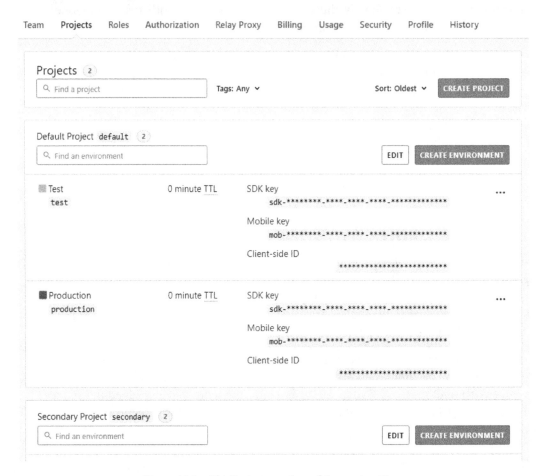

Figure 13.3 – The Projects section of Account settings

In *Figure 13.3*, there are two projects shown, **Default Project** and **Secondary Project**. As a LaunchDarkly account might have many projects set up, it is possible to filter the listed projects by searching for their names or via filtering through the tags that are added to a project. It is possible to sort the displayed projects by when they were created or by their name. Clicking on the **CREATE PROJECT** button reveals a new screen in which you only need to enter a few pieces of information:

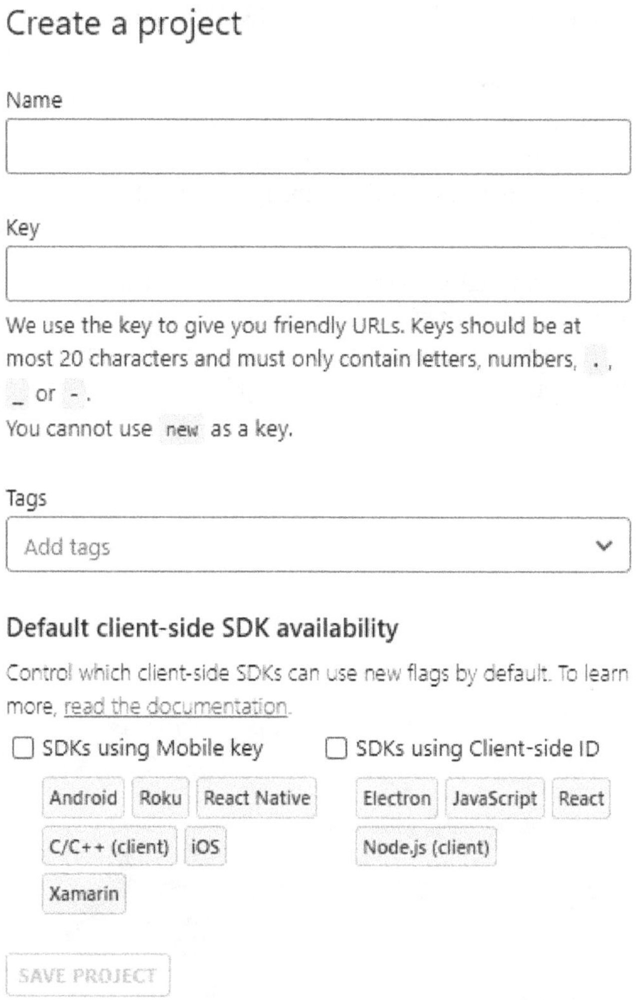

Create a project

Name

[   ]

Key

[   ]

We use the key to give you friendly URLs. Keys should be at
most 20 characters and must only contain letters, numbers, . ,
_ or - .
You cannot use new as a key.

Tags

[Add tags   ⌄]

Default client-side SDK availability

Control which client-side SDKs can use new flags by default. To learn
more, read the documentation.

☐ SDKs using Mobile key   ☐ SDKs using Client-side ID

    Android Roku React Native   Electron JavaScript React

    C/C++ (client) iOS   Node.js (client)

    Xamarin

[SAVE PROJECT]

Figure 13.4 – The Create a project screen

A project requires a **Name,** and just as with creating a feature flag, the name defaults as the
Key for the project, although it is possible to change the **Key** to something else if needed.
At the point of creating a project, **Tags** can be added. **Tags** help with project management,
which is especially useful when using projects for key customer experiences and user
journeys rather than a team-based approach. The tags can help show you which experiences
and/or products are contained within the project. The last part of the setup process is to
select whether the flags in this project will be using mobile and/or client-side SDKs. All of
this information, apart from the **Key**, can be changed when needed via the **Edit** button of
the project. Deleting a project is available from the **Edit** button too, but you need to be in
a different project to delete one. You cannot delete the project that you are currently in.

Once a project has been created, by default, it will have a **Test** environment and a **Production** environment. These are shown in LaunchDarkly with orange- and green-colored boxes next to their names. The **Projects** section displays the environment's **Time to Live** (**TTL**), which refers to how long the settings of the environment are cached, and the SDK keys. In *Figure 13.3*, the keys are redacted; when you look at your own environments, the keys are visible and can easily be copied to be used within your applications to initialize the LaunchDarkly SDK client.

Environment and approval settings are available from the three dots menu on the right-hand side of each environment. The environment settings deal with the appearance and configuration of the project in LaunchDarkly; the approval settings are for how changes to the flags in the environment can be made. First, we will take a look at the environment settings:

Edit Test environment

Name

Test

Tags

Add tags

☐ Enable secure mode

☐ Require comments for flag and segment changes

☐ Require confirmation for flag and segment changes

☐ Send detailed events to data export destinations.

TTL (0-60 minutes)

0

Color

#4c4536

SAVE ENVIRONMENT

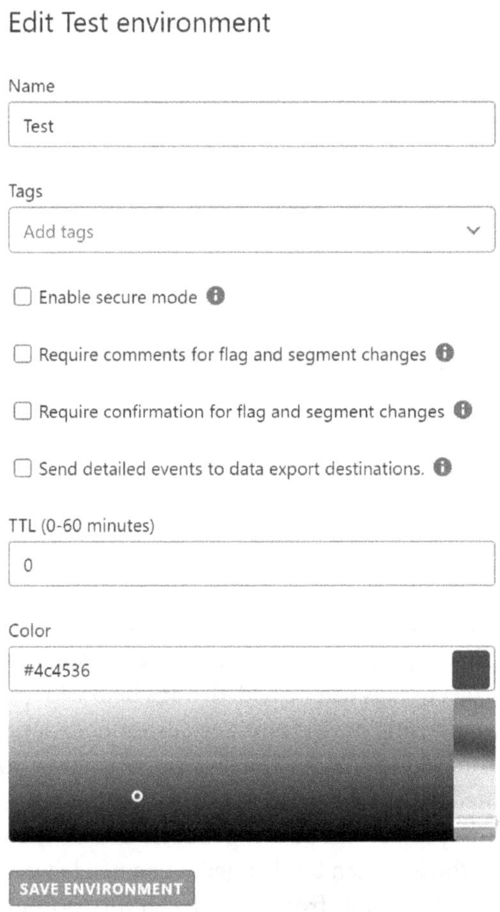

Figure 13.5 – The first part of the environment settings

As with projects, each environment can have a **Name** and **Tags**. Here, the tags can be used to indicate the purpose of the environment. Apart from the production environment, most other environments are going to be used for several types of testing, and **Tags** can help indicate the various types of testing carried out on each environment (unless, of course, the view is that all testing can be done in production and then separate test environments might not be needed). Next, there are four checkboxes to set environment-wide changes:

- **Enable secure mode**: This setting is used when a client-side SDK is in use. It ensures that users cannot impersonate other users when the SDK is identifying the user. When using the client-side SDK, this is something that should be considered for use in production. If this is needed in production, then some or all test environments might need it enabled to ensure that the same experience is available when testing functionality.

- **Require comments for flag and segment changes**: This setting is geared toward improving the governance and change management processes of LaunchDarkly. When enabled, a comment must be added to all changes for feature flags and segments. This helps provide context as to why the change was made. It can also help to prevent any accidental changes, as people are required to do something extra before saving a change. This environment-wide configuration is only available with an enterprise plan.

- **Require confirmation for flag and segment changes**: Again, this setting helps with the usage of LaunchDarkly itself. The confirmation that must be provided is to type in the **Name** or **Key** of the flag. This is particularly useful to reduce any accidental changes, as people have to check that the flag they are working on is the right flag. Therefore, it reduces the chance of making changes to the wrong flag.

- **Send detailed events to data export destinations**: If data export is being used for flags within this environment, it is possible to turn it on or off for all flags. This can save time as there is no need to go through each flag individually to turn this feature on.

The next setting is the **TTL**, which can be provided with a number between 0 and 60 to indicate how long the settings are saved. Setting longer times here will result in fewer checks from the LaunchDarkly SDK back to the LaunchDarkly service but will result in any changes made taking longer to appear within the applications.

Next is the opportunity to change the color of the environment. This color appears in the upper-left corner of LaunchDarkly where the project and environment information is shown. The visual indication of which environment someone is in is extremely useful to help people make changes in the correct one. It is worth having a common approach to environment colors so that people who might change flags across projects do not make mistakes based on the environment's color. By default, the test environment is orange and the production environment is green. It is recommended that you keep to this approach and ensure all the production environments across your projects are green in color. Other colors could be used for different testing purposes, but green should probably be kept just for production.

The last couple of aspects of the environment settings screen are the most important, as shown here:

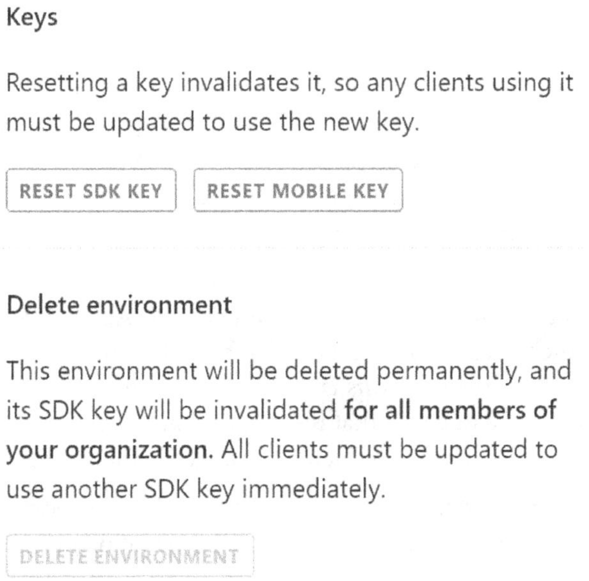

Keys

Resetting a key invalidates it, so any clients using it must be updated to use the new key.

[RESET SDK KEY]   [RESET MOBILE KEY]

Delete environment

This environment will be deleted permanently, and its SDK key will be invalidated **for all members of your organization.** All clients must be updated to use another SDK key immediately.

[DELETE ENVIRONMENT]

Figure 13.6 – The second part of the environment settings

First is the ability to reset the environment's SDK keys. This is important if the keys have somehow been compromised, which would allow other people and applications to access your LaunchDarkly environment via an SDK. Resetting the SDK keys is, therefore, not something that should be done lightly, but it is necessary to have this functionality. The last part of the settings screen is the option to **DELETE ENVIRONMENT** This results in the immediate invalidation of the SDK keys, and any applications that are using them will no longer be able to reach LaunchDarkly. It is worth checking the usage of feature flags for an environment before deleting it as the SDK keys would need to be replaced in all applications, which might not be a quick or easy exercise if this were done accidentally.

Some of these options have a significant impact, and so managing these changes is important, as is managing the permissions of the members of your team. In the next section, we will explore the roles and permissions available in LaunchDarkly to help mitigate and control these risks.

Learning about roles and permission management

Throughout this book, I have emphasized how being able to make changes easily to production through feature management is an immensely powerful option for businesses in order to manage and fine-tune their products. However, it comes with the risk that if changes are made accidentally or maliciously, then the impact could be substantial. To reduce this risk and to make sure the right individuals have the ability to make changes to the right flags and on the right environments, roles and permissions are used within LaunchDarkly.

There are three built-in roles that LaunchDarkly offers to provide a basic set of permission rules. These default roles are as follows:

- **Reader**: Members of the LaunchDarkly account with this role can only access and view LaunchDarkly but are unable to make any changes. This might be useful for stakeholders or members of support teams who would benefit from being able to see the targeting rules and configuration of a flag but are not expected to make any changes to them.

- **Writer**: This role allows individuals to make changes to the main functionality of LaunchDarkly but not the administration of the account. Making changes to feature flags, segments, metrics, projects, and a few other aspects are all allowed, but changing the membership settings, billing, or account-wide functionality is all restricted. This is good for most people on the team who should be able to amend the behavior of applications and the data recorded but shouldn't be able to change the setup of LaunchDarkly.

- **Admin**: The admin role has the ability to change all aspects of LaunchDarkly from the account-wide settings to individual feature flags. However, there is one thing that general admin members cannot do and that is changing the owner of the LaunchDarkly account. That functionality is only available to the owner account. Few people should be set as admins, and they should only make changes to **Account settings** when necessary, although they might make frequent changes to flags and segments.

These default roles work well for a small number of people within a LaunchDarkly account; however, when multiple teams or even departments start working in one account, there is a need to extend and customize these roles. This is where the **Custom roles** come in. The custom roles feature is only available with an enterprise plan, but it is worth exploring to understand the powerful configuration options that it provides.

In the **Roles** section of **Account settings**, a list of all the configured custom roles is shown:

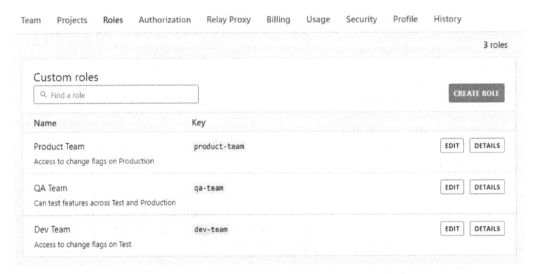

Figure 13.7 – The Roles section of Account settings

This list of custom roles can be searched to find specific roles. This is very useful as the number of members grows and there is a need for bespoke permissions. From this list, it is possible to view the name, description, and key. As with feature flags themselves, the **Name** and description are both editable, but the **Key** is not. The permissions set on the role are also editable, so roles can easily be changed as needed over time. Clicking on **CREATE ROLE** will take you to a new screen where you can set up a new role:

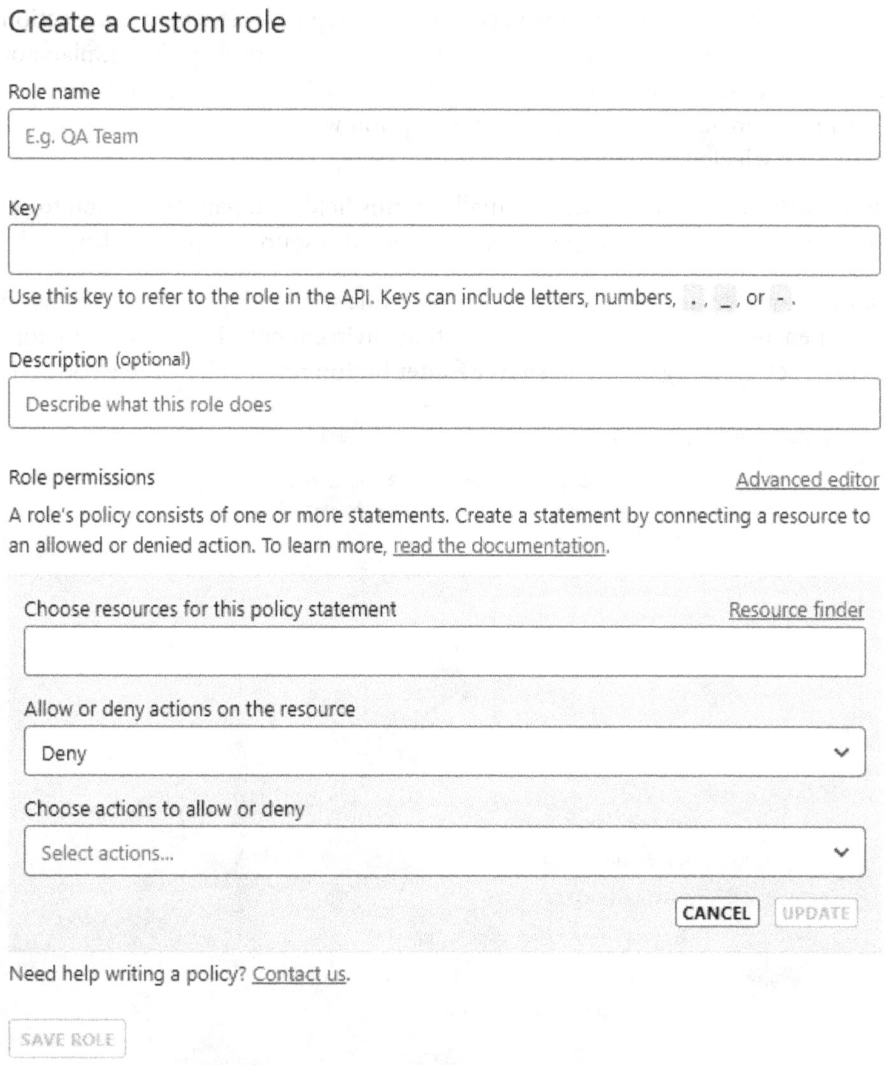

Create a custom role

Role name

```
E.g. QA Team
```

Key

```

```

Use this key to refer to the role in the API. Keys can include letters, numbers, ., _, or -.

Description (optional)

```
Describe what this role does
```

Role permissions   Advanced editor

A role's policy consists of one or more statements. Create a statement by connecting a resource to an allowed or denied action. To learn more, read the documentation.

Choose resources for this policy statement   Resource finder

```

```

Allow or deny actions on the resource

```
Deny                                                              ⌄
```

Choose actions to allow or deny

```
Select actions...                                                 ⌄
```

[CANCEL] [UPDATE]

Need help writing a policy? Contact us.

[SAVE ROLE]

Figure 13.8 – Creating a new custom role

The first three fields are self-explanatory when creating a custom role: **Role name**, **Key**, and **Description**. They are just like what we have viewed elsewhere in this book: feature flags, segments, and metrics. The **Role permissions** section is where the real value of the role functionality comes in. There are two views for setting the permissions, and to begin with, we will explore the simple editor before we go on to examine the **advanced editor**. There are three steps to creating a rule:

1. **Choose resources for this policy statement**: In this field, you select the resources to which the rule will apply.

2. **Allow or deny action on the resource**: The next step is to set whether the action will be an **allow** action or a **deny** action. These terms are perhaps self-explanatory, but to clarify, the **allow** option is used to enable those members in a role to perform certain actions, whereas the **deny** option will restrict them from conducting selected actions.

3. **Choose actions to allow or deny**: Finally, in this field, you have the option to set the action against the individual pieces of the selected resource's functionality.

In this example, I will create the rules for a developer role where they will have permission to update the test environments but not the production environments. First, a resource for the rule needs to be set. Clicking on the **Resource finder** button reveals all possible resources:

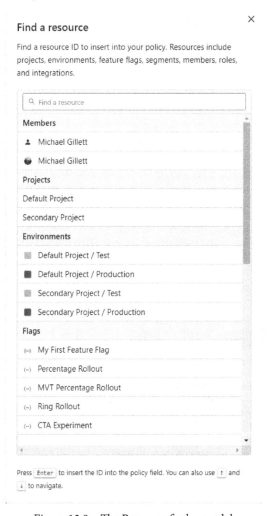

Figure 13.9 – The Resource finder modal

The preceding list of resources can be searched. This is useful in large LaunchDarkly accounts where many resources exist. It is possible to navigate through the list, and it is grouped into the following:

- **Members**
- **Projects**
- **Environments**
- **Flags**
- **Integrations**
- **Segments**
- **Metrics**
- **Roles**

Clicking on one of the resources will close the popup and fill the resource field. Depending on which type of resource has been selected, the value in the field might not be immediately obvious. This is because it is showing the technical name and syntax of the resource. LaunchDarkly does provide a human-readable definition of the value, too:

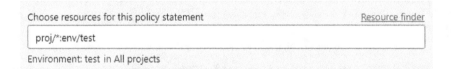

Figure 13.10 – The resource selected for the role permissions

In *Figure 13.10*, the resource of a test environment was selected. This has resulted in all of the environments named **test** being the target for this rule. This rule goes for all projects as it is specifically looking at the environment name and ignores which project it is in.

Next, the action to be applied to the resource needs to be configured. There are only two options for this: **allow** or **deny**. In this example, **allow** is being selected. The last step is to select which aspects of the environment are going to be allowed. The drop-down menu for an environment is quite long since there is a lot that can be specified:

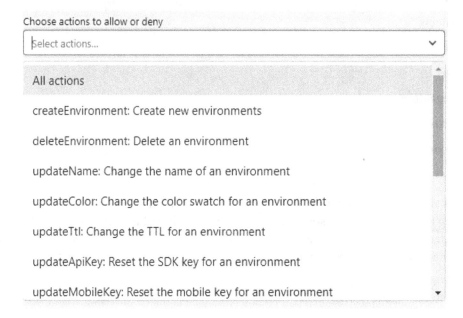

Figure 13.11 – The actions that can be allowed or denied for an environment

In this example, the **All actions** option is selected. Clicking on the **Update** button reveals a new UI within the screen to create a new role:

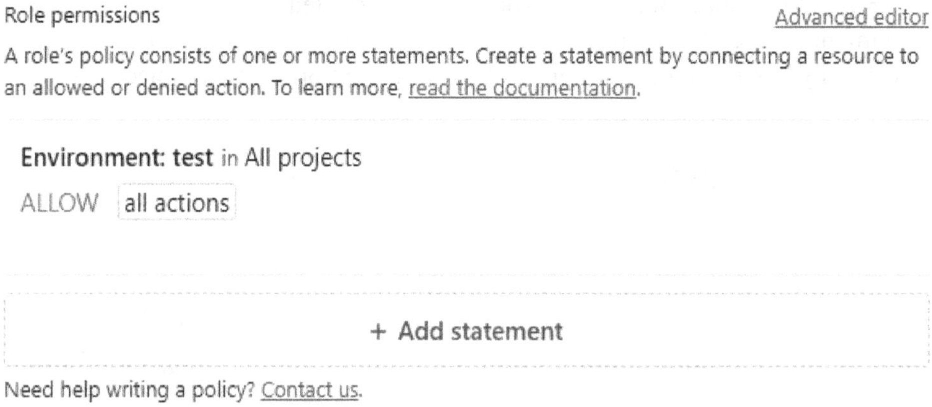

Role permissions   Advanced editor

A role's policy consists of one or more statements. Create a statement by connecting a resource to an allowed or denied action. To learn more, read the documentation.

Environment: test in All projects
ALLOW   all actions

+ Add statement

Need help writing a policy? Contact us.

Figure 13.12 – A configured permission rule on a custom role

The new UI displays an easy-to-read version of the rule. From this, you can quickly tell that this rule is targeting all test environments across all projects and that all actions are being allowed. There is an **Add statement** button that can be used to add more statements to the permissions. Clicking on this reveals the same role permissions UI that we just looked at. To complete the example, a second rule is needed that will deny access for this role to update the production environments. This is almost identical to the previous steps except that this will be a rule against all environments named **production**, and the action will be **deny**:

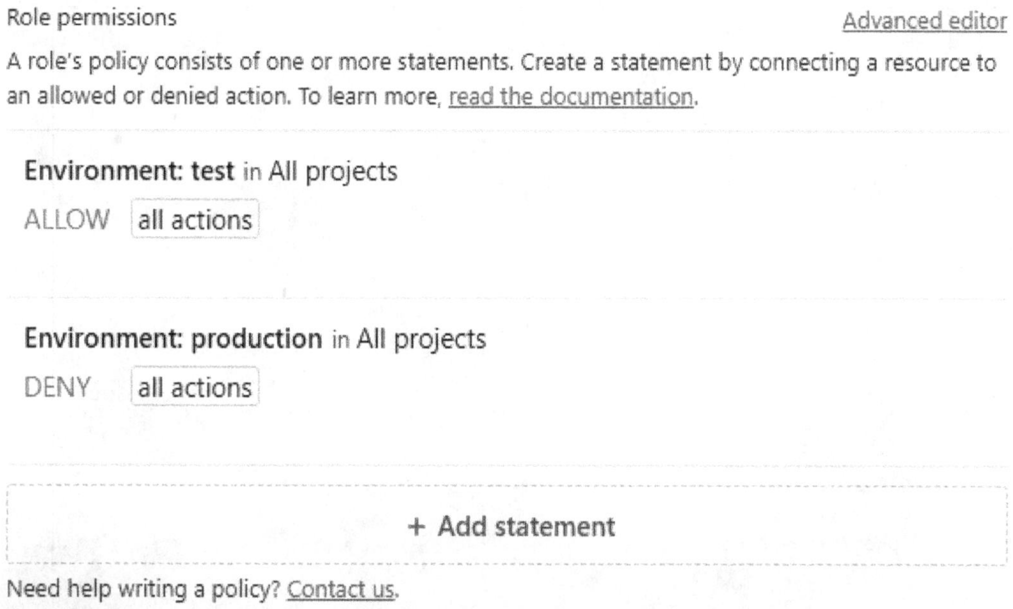

Role permissions   Advanced editor

A role's policy consists of one or more statements. Create a statement by connecting a resource to an allowed or denied action. To learn more, read the documentation.

Environment: test in All projects

ALLOW   all actions

Environment: production in All projects

DENY   all actions

+ Add statement

Need help writing a policy? Contact us.

Figure 13.13 – The completed rules to allow changes to the test environments but not production

This role permission is now configured to both enable and disable users from accessing certain parts of LaunchDarkly. With the granularity of the resources that can be selected and the actions that can be performed, any type of permissions can be created. So far, the UI that has been shown might not be the most efficient way of setting up complex rules, which is why there is also an **advanced editor**:

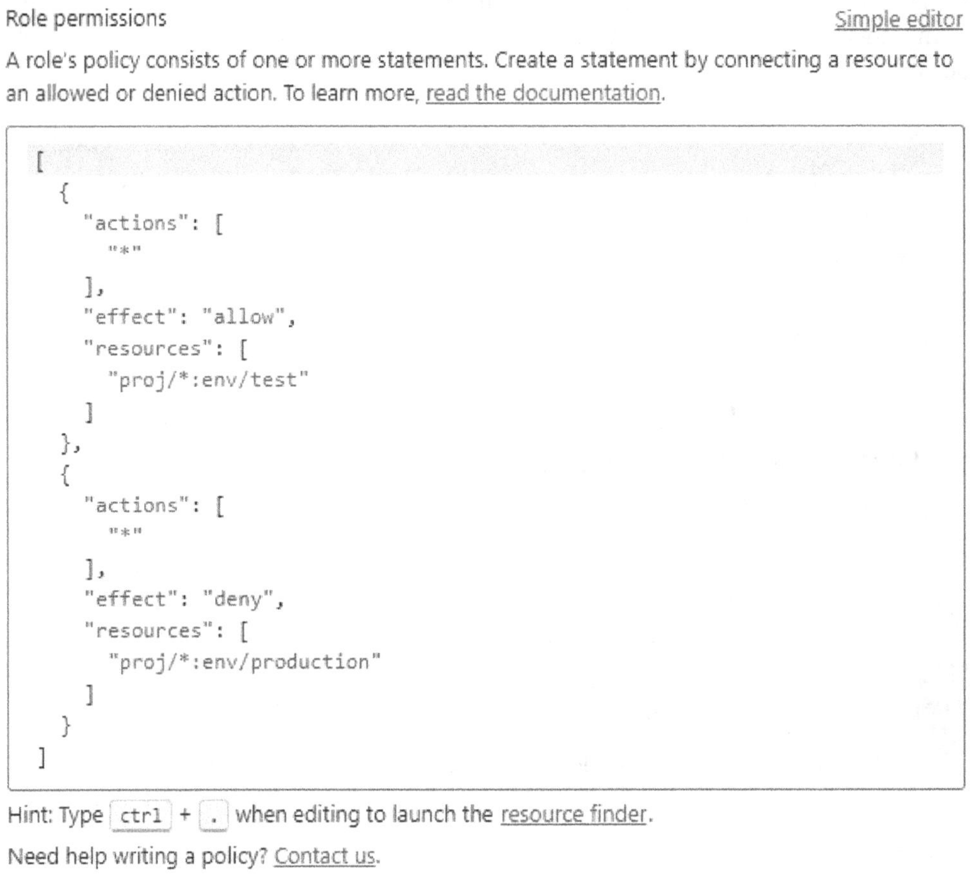

Role permissions   Simple editor

A role's policy consists of one or more statements. Create a statement by connecting a resource to an allowed or denied action. To learn more, read the documentation.

```
[
  {
    "actions": [
      "*"
    ],
    "effect": "allow",
    "resources": [
      "proj/*:env/test"
    ]
  },
  {
    "actions": [
      "*"
    ],
    "effect": "deny",
    "resources": [
      "proj/*:env/production"
    ]
  }
]
```

Hint: Type ctrl + . when editing to launch the resource finder.

Need help writing a policy? Contact us.

Figure 13.14 – The advanced editor for role permissions

The advanced editor reveals the JSON object that is used for the role permissions. This is actually what is used even when using the simple editor—the UI just masks it. Within this editor, there is full control over what is being specified, which can help you to create the list of actions or resources more quickly. There are some syntax elements to be aware of such as * being used as the wild card. In *Figure 13.14*, this is used in both `actions` to show that all actions are within the scope of the rule and also in `resources` to show that all projects are in scope.

Of course, using this experience is not as straightforward as the simple one, but to look up the list of resources, the **resource finder** is still available via a keyboard shortcut: *Ctrl + ..*

Some example roles and permissions could be as follows:

- **Team restrictions**: This is where people within a certain team can only access certain flags or flags within certain projects.

- **Role restrictions**: This is where depending, on a person's role within the team, they are given access to some environments. This is effectively the previous example.

- **Specific flags**: This is where people are given access to change specific flags. This could be something done for stakeholders, or in the scenario of kill switches, it could be something to which members of the support team are given access.

In reality, there are endless possibilities for how permissions can be configured. With this powerful system, businesses and teams can define how best to manage people working with feature management. One thing to note is that if there are too many barriers in place for teams to work effectively with feature management, then some of the advantages can be lost, especially as feature management is an extension of good DevOps practice. Just like a mature DevOps software development team can control all aspects of their production environment, the same should be true when it comes to having access to making production changes within LaunchDarkly.

Once the roles have been created, members need to be assigned to them, as shown in *Figure 13.2*.

Ensuring good role and permission management goes a long way to ensuring that your LaunchDarkly account is well controlled. However, another key area is ensuring that only the right people can log into the system. In the next section, we will explore how people and applications are authenticated within LaunchDarkly.

Learning about authorization and security

The next two sections of **Account settings** that will be covered are **Authorization** and **Security**. These are two separate sections on the settings page, but both relate in terms of ensuring that your LaunchDarkly account and the access to it are secure. Let's begin with the **Authorization** section:

Account settings

Team	Projects	Roles	**Authorization**	Relay Proxy	Billing	Usage	Security	Profile	History

Authorized applications

Applications you or your team have authorized to access LaunchDarkly. To start developing an OAuth application, contact support@launchdarkly.com. Read Authorizing OAuth Applications.

Q Find an application	Authorized by: Anyone ⌄

There are no authorized applications yet.

1 access token

Access tokens

Access tokens are used to authenticate with our REST API. Treat your tokens as sensitive information — only you can see the values of tokens you create.

Q Find an access token	Created by: Anyone ⌄	CREATE TOKEN

Description	Role(s)	Last used	Creator	
Code References CLI api-****1d73	Writer	2 months ago	Created by you	...

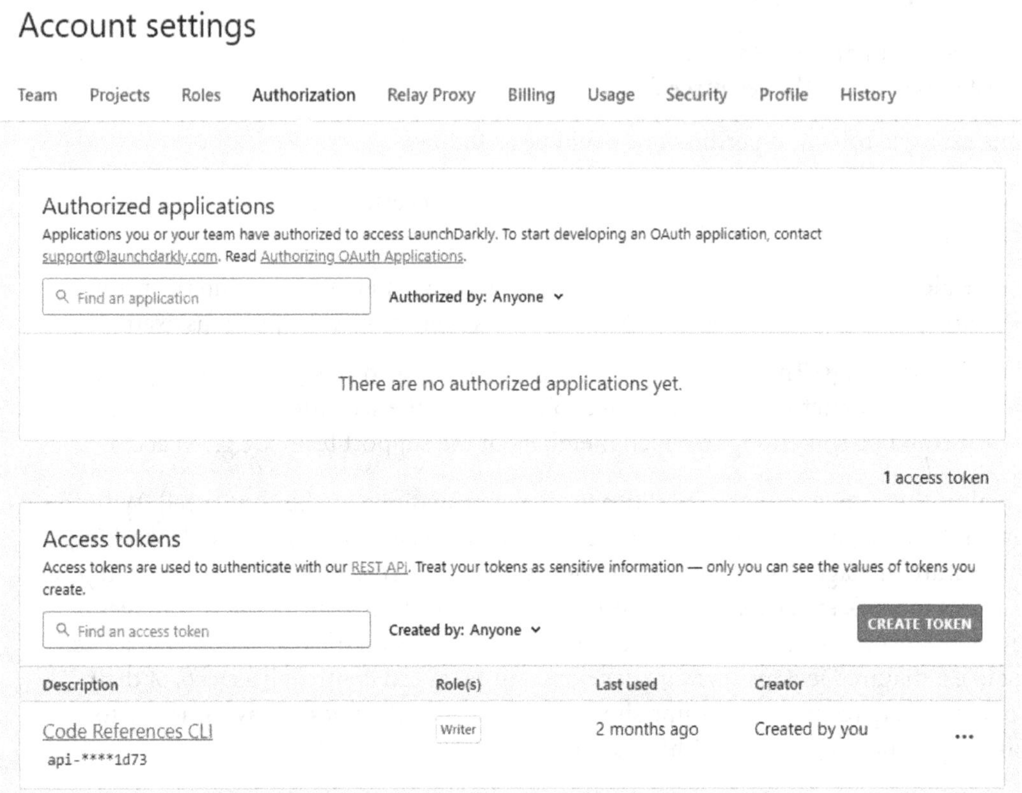

Figure 13.15 – The Authorization section of Account settings

There are two aspects to this section, as follows:

- **Authorized applications**: These are the applications that connect to this LaunchDarkly account through an OAuth connection. They are used for integrations with other tools, such as Slack or Microsoft Teams. This type of integration can be useful for receiving notifications or as a trigger when key events occur in LaunchDarkly. For example, a team might want notifications when a new flag is created or when targeting rules are changed in production flags. This helps a team stay aware of how their production applications are configured.

- **Access tokens**: In this second part, you can find the applications that need access tokens to make use of LaunchDarkly's REST API. One example of an application that needs an access token is the code references tool that was detailed in the *How can LaunchDarkly help with trunk-based development* section of *Chapter 7, Trunk-Based Development*. These access tokens need to be kept secret, as they enable applications to read a lot of information and make changes to the setup of flags.

Clicking on the **CREATE TOKEN** button will reveal a new screen for making new access tokens:

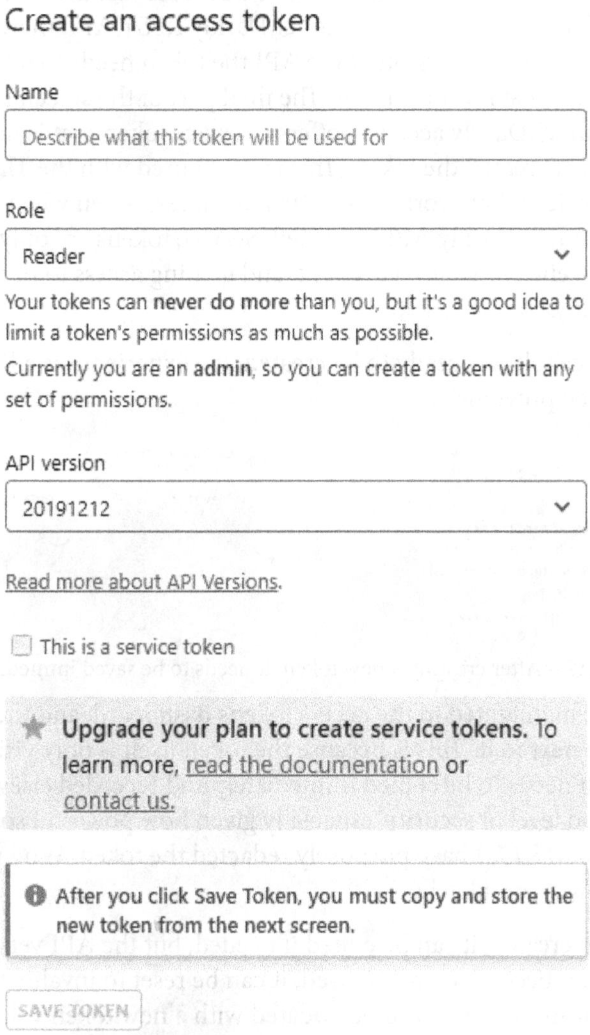

Create an access token

Name

```
Describe what this token will be used for
```

Role

```
Reader                                                    ⌄
```

Your tokens can **never do more** than you, but it's a good idea to limit a token's permissions as much as possible.

Currently you are an **admin**, so you can create a token with any set of permissions.

API version

```
20191212                                                  ⌄
```

Read more about API Versions.

☐ This is a service token

⭐ Upgrade your plan to create service tokens. To learn more, read the documentation or contact us.

ℹ After you click Save Token, you must copy and store the new token from the next screen.

```
SAVE TOKEN
```

Figure 13.16 – Creating a new access token

Each access token needs a **Name**, and it is recommended that you are descriptive about the purpose of the token. Since tokens are powerful, they need to be well managed, and a good naming convention can help convey their use and the associated level of access that each token has. This information can stop accidental duplicate tokens from being created that have the same access rights but different names.

Setting the **Role** is the crucial aspect in defining the permissions that the token has. The list of roles includes all the roles set up within LaunchDarkly, so a token can have the same set of permissions as any member. However, there is also the option for an **inline policy** that allows for unique permissions to be set for just the access token. Given that tokens are a mechanism to access the LaunchDarkly REST API, there is an **API Version** option to specify which version of the API the token needs to integrate with. By default, this is set to the most recent version. The final part of the screen is an option that, when enabled on a LaunchDarkly account, offers a service token that is not tied to the individual's account that created the token. This is configured with the **This is a service token** checkbox. A service token works better than an access token when a long-lived integration with the LaunchDarkly API is needed. Service tokens are only available as an extra component to an enterprise-level account, and gaining access to this feature requires talking to LaunchDarkly.

Once access or a service token is ready to be created, the experience is a little different from what we have seen previously:

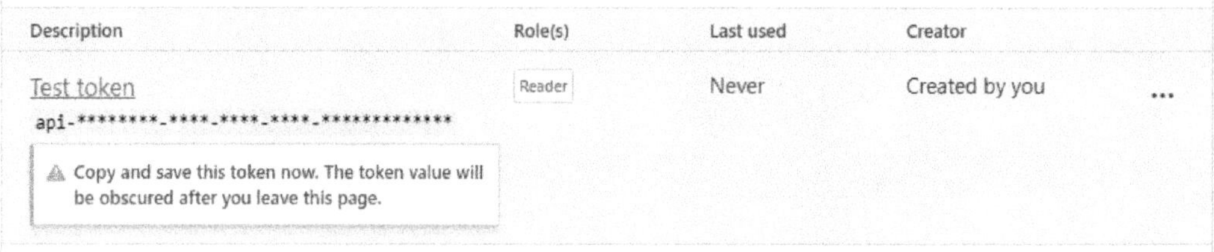

Description	Role(s)	Last used	Creator	
Test token api-********_****_****_****_************ ⚠ Copy and save this token now. The token value will be obscured after you leave this page.	Reader	Never	Created by you	...

Figure 13.17 – After creating a new token, it needs to be saved immediately

The new token will be highlighted in the access tokens dashboard, and there will be a warning message next to it. This is because the token itself is only visible in LaunchDarkly once. It needs to be copied immediately and recorded elsewhere. This practice ensures a good level of security, especially given how powerful some of the access tokens can be. In *Figure 13.17*, I have purposely redacted the token. When you create one, you will be able to view it completely.

After a token has been created, it can be edited if needed, but the API version cannot be changed. If a token does become compromised, it can be reset to invalidate it, and this will require all of the applications using it to be updated with a new token value. Additionally, tokens can be cloned to speed up the creation of similar tokens. Finally, tokens can be deleted. All this functionality is available in the three dots menu of the tokens.

In terms of making sure that good security is adopted across all aspects of LaunchDarkly, there are some options available within the **Security** section of **Account settings**:

Account settings

Team   Projects   Roles   Authorization   Relay Proxy   Billing   Usage   **Security**   Profile   History

Multi-factor authentication

Requiring an additional authentication method adds another level of security for your organization.

☐ Require multi-factor authentication for new members

`SAVE`

Team management

Manage authentication via SSO and SCIM. To learn more, read the documentation.

Single sign-on status: **Not configured**
Team members sign in via their email and password.

`CONFIGURE SAML`

SCIM provisioning status: **SAML not enabled**
No user provisioning for your account.

Session configuration

Session duration

| 14 | Days ⌄ |

☐ Refresh sessions automatically ⓘ

`SAVE`

Revoke all sessions

This will log all users out of LaunchDarkly and force them to re-authenticate.

`REVOKE ALL SESSIONS`

Enable enhanced support

Allow LaunchDarkly's support team to temporarily view your account to help diagnose and resolve your support request. The support team will only have view access to your account if this feature is enabled and their access will automatically expire after the date/time you've indicated when enabling this feature.

`ENABLE`

`SAVE`

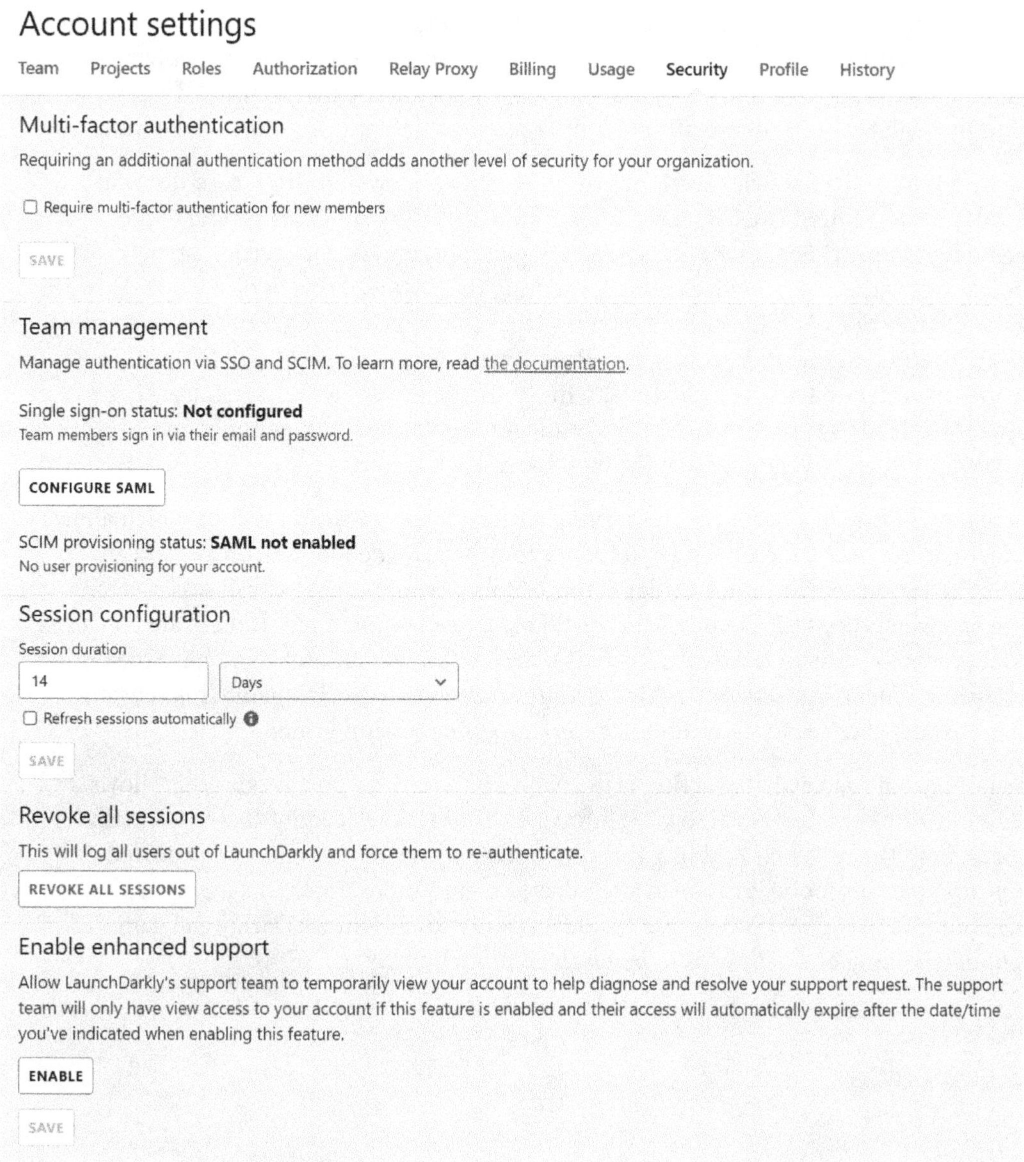

Figure 13.18 – The Security section of Account settings

The first part of the **Security** section is to enable MFA for all new members of this LaunchDarkly account. This is something that is triggered at the point of account creation, so it can easily be applied to new accounts. However, for existing accounts, it needs to be triggered for individual members who do not have it enabled. *Figure 13.2* shows how MFA can be enabled for a single account. If users only use an email address and password to log in to LaunchDarkly, it would be advisable to enable MFA for extra security. If members log in through SSO or other forms of identity providers, then this might not be necessary.

The next part of the **Security** section is to set up SSO and **System for Cross-domain Identity Management** (**SCIM**) integrations. LaunchDarkly supports several of the larger identity providers, including Okta, Google Apps, and Azure. Setting up new integrations varies per provider and is something that is out of this book's scope. However, LaunchDarkly does provide documentation regarding this topic. One thing worth considering is that if your business uses **Role-Based Access Control** (**RBAC**) within your identity provider systems, it is possible to integrate these roles with LaunchDarkly's roles. These roles need to exist in LaunchDarkly already, but the assignment of people to roles can be done via RBAC rather than through the tool's UI.

There is an option to specify the duration of a user's session within the tool, which allows greater control over a member's session once they have logged in. The units that can be selected are minutes, hours, or days. There is also the option to refresh sessions for someone when they visit LaunchDarkly with an active session open. If there are concerns about access to LaunchDarkly, and there is a need to log out all users, you have the option to revoke all members' sessions, which requires everyone to log in again. This is useful when making changes to the authentication process or session limits.

The last part of the **Security** section is to enable enhanced support mode. This allows LaunchDarkly's support team to view information about the account that is normally hidden from them. When enabling this functionality, an end date must be provided to ensure that this mode is disabled after a certain amount of time. This is helpful when additional support is required to address an issue, as LaunchDarkly can gain an understanding of what needs to be resolved more quickly.

Through the various options, it is possible to implement measures to ensure that your LaunchDarkly account is well secured. In the next section, we explore the **Billing** and **Usage** sections of the tool.

Understanding billing and usage

The next sections that will be covered in **Account settings** are **Billing** and **Usage**. These two sections go hand in hand. **Billing** provides information about the type of plan that the account is using, and the **Usage** section shows how much of the quota of the account's plan remain from the previous month.

There are four metrics that determine the usage of a plan:

- **Seats**: This is the limit of how many members will be in the account.

- **Client-side MAUs**: This is the limit of how many **Monthly Active Users** (**MAUs**) there will be from the client-side SDK. This forms part of the plan as the evaluation is done on LaunchDarkly's hardware, and they incur this cost and that of the bandwidth requirements. When using server-side SDKs, the evaluation is done by your application on your own hardware, so you are paying that cost. The fact that the usage of the client-side SDKs can have an impact on what plan is needed and how much LaunchDarkly will cost is ultimately why, throughout this book, I have advised against using that version of the SDK.

- **Experimentation events**: This is the limit for how many experimentation events can be performed per month. These events come from the track metric event functionality that was explored in *Chapter 11*, *Experiments*.

- **Data export events**: This is the limit for how many data export events can happen per month. Data export events have only been briefly mentioned in this book, but they can be used to integrate LaunchDarkly's collected data with other systems; however, this can come at a cost.

The **Billing** section shows what the limits are for the account along with the progress that has been made through the limits at which the account currently stands:

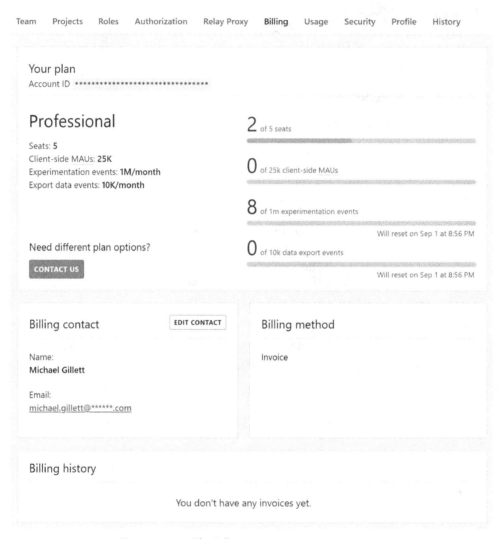

Account settings

Team   Projects   Roles   Authorization   Relay Proxy   **Billing**   Usage   Security   Profile   History

Your plan
Account ID ********************************

Professional

Seats: **5**
Client-side MAUs: **25K**
Experimentation events: **1M/month**
Export data events: **10K/month**

Need different plan options?

CONTACT US

2 of 5 seats

0 of 25k client-side MAUs

8 of 1m experimentation events

Will reset on Sep 1 at 8:56 PM

0 of 10k data export events

Will reset on Sep 1 at 8:56 PM

Billing contact   EDIT CONTACT

Name:
Michael Gillett

Email:
michael.gillett@******.com

Billing method

Invoice

Billing history

You don't have any invoices yet.

Figure 13.19 – The Billing section in Account settings

The top of the page shows the four quotas that comprise the makeup of the account plan. The limits are shown on the left, while a visual indicator of the monthly progress through those limits is shown on the right. For any metrics that are reaching their limits, there are two options: change the code of the application to trigger fewer events or speak to LaunchDarkly about increasing the limits.

The remainder of this section shows the financial aspects of the account with information about the billing contact, the method of payment, and any previous invoices. With a legitimate account as opposed to my example account, there would be more useful information on this page.

Where there are limits that are being approached by the client-side MAUs, experimentation events, or data export events, more information can be gained about their cause from the **Usage** section:

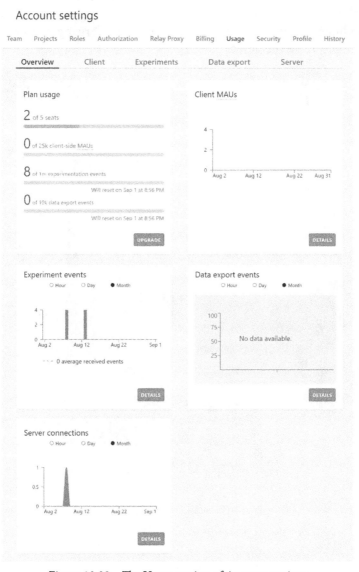

Figure 13.20 – The Usage section of Account settings

The **Usage** page shows the plan's metric, apart from the seat count, along with information about the number of server connections. While this does not form part of the billing model, it is interesting data that shows how the integration with LaunchDarkly and the usage of the tool change over time. Across the top of this section, there are five tabs: **Overview**, **Client**, **Experiments**, **Data export**, and **Server**. The **Overview** tab shows information about all the metrics along with the same UI of the plan's usage that is in the **Billing** section. This overview screen can quickly show off any metrics that have changed significantly over the past four weeks. If more detailed information is needed, or if viewing a different date would be more useful, then clicking on the **DETAILS** button or the corresponding tab at the top of the page will present a larger graph with more controls. For example, the **Experiments** events graph appears as follows:

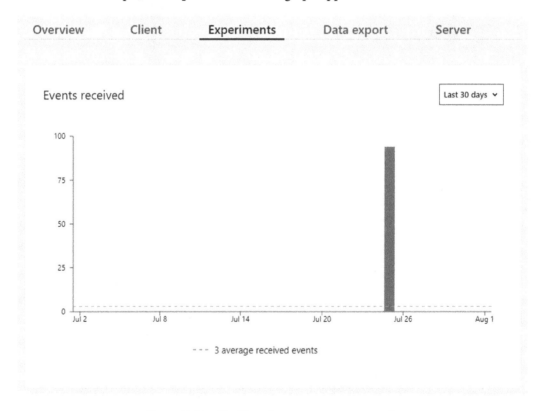

Figure 13.21 – The Experiments events usage graph

The preceding graph shows the average number of events over the time frame with a dotted line. Also, it shows the daily total of events recorded. There was only one day where experiment metrics were being triggered, so it stands out. The time frame can be refined to a maximum of 60 days or down to the last 60 minutes.

In normal usage, these graphs might show some trends, but anything that deviates from the norm could be down to a new implementation or misconfiguration. It is good practice to regularly review all this data, especially following any major changes to the LaunchDarkly implementation in an application, to ensure that the usage remains within the limits.

In the **Account settings** section, not only are there settings that affect all members along with the billing information, but there is also the functionality to change your own profile. We will explore that next.

Exploring your profile

The penultimate section within the **Account settings** section is the **Profile** section. On this page, LaunchDarkly shows what roles a user has and allows someone to change their name and email address. The email address is used to log in to their account, so if an identity provider has been integrated, then changing the email address is not possible.

The user's password can also be changed from this section. The last part of this page is to show the MFA status of the user's profile. It is possible for the individual to enable MFA on their account from here, rather than having an admin trigger it, as we explained earlier in this chapter.

Discovering history

The concluding section to examine is the **History** one. Previously, we looked at the history of individual feature flags and the audit log for a project and environment, and the **Account settings History** page is similar to these. The information contained within this log is everything that impacts the LaunchDarkly account. It is possible to search the account history for events and refine the date range of the displayed logs. For any individual event in the log, there is a **DETAILS** button that shows detailed information of what occurred. This includes the JSON of the changes made, just as we demonstrated in *Figure 12.11*, in *Chapter 12, Debugger and Audit Log*.

Summary

To get the most out of feature management, there needs to be a certain level of confidence and reliability in the tool that is being used to change the availability of functionality within a production system. LaunchDarkly not only provides the functionality to enable feature flags targeting, experimentation, and rollouts, but it also offers a wealth of other features to manage the risk of inadvertent changes to production or malicious actors getting into the system. This chapter, although not dealing with feature management functionality directly, shows how LaunchDarkly can be configured according to the team and business needs.

Using projects and environments, flags can be grouped to best serve different team structures and business needs. Measures can be imposed on certain environments to ensure good governance and to limit the chance of negative changes being made. Adding to that, members of the LaunchDarkly account can be limited in what they can do using the powerful roles and permissions system. Teams or individuals can be empowered or blocked from making changes based on their need to be able to access parts of LaunchDarkly.

By using MFA and SSO, there can be good security measures put in place to ensure that bad actors cannot easily get into the system. This also extends to the use of access and service tokens so that automations and integrations offer the same robust access controls.

This chapter also looked at the billing and usage information provided by LaunchDarkly for the account. Keeping an eye on this information ensures that costs are well managed and that any misconfigurations, or unexpected implementations, are addressed quickly before they become expensive for the business.

Finally, this chapter briefly looked at the options for profile management and the account's history log. The level of information contained within the log is similar to what we discussed in earlier chapters and, again, shows how all changes within LaunchDarkly are captured for future review. This helps to build confidence in the tool.

This closing chapter completes the journey of discovery into feature management with LaunchDarkly. From introducing the concept of feature flags, both temporary and permanent, to understanding how to gain insights from experiments, there have been many examples of how LaunchDarkly can enhance the modern DevOps approach to build, deploy, test, and deliver software.

The information provided within this chapter shows how LaunchDarkly is a powerful and mature tool that can be used to help teams manage feature flags. This is in addition to all the other functionality detailed in this book that LaunchDarkly provides for targeting features to individual customers, segments, and any rules needed based on custom attributes. With such powerful and flexible systems, LaunchDarkly makes it easy for teams to truly take advantage of feature management and deliver new software to their customers quickly and safely.

Packt.com

Subscribe to our online digital library for full access to over 7,000 books and videos, as well as industry leading tools to help you plan your personal development and advance your career. For more information, please visit our website.

Why subscribe?

- Spend less time learning and more time coding with practical eBooks and Videos from over 4,000 industry professionals

- Improve your learning with Skill Plans built especially for you

- Get a free eBook or video every month

- Fully searchable for easy access to vital information

- Copy and paste, print, and bookmark content

Did you know that Packt offers eBook versions of every book published, with PDF and ePub files available? You can upgrade to the eBook version at packt.com and as a print book customer, you are entitled to a discount on the eBook copy. Get in touch with us at customercare@packtpub.com for more details.

At www.packt.com, you can also read a collection of free technical articles, sign up for a range of free newsletters, and receive exclusive discounts and offers on Packt books and eBooks.

Other Books You May Enjoy

If you enjoyed this book, you may be interested in these other books by Packt:

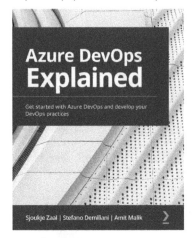

Azure DevOps Explained

Sjoukje Zaal, Stefano Demiliani, Amit Malik

ISBN: 9781800563513

- Get to grips with Azure DevOps
- Find out about project management with Azure Boards
- Understand source code management with Azure Repos
- Build and release pipelines
- Run quality tests in build pipelines
- Use artifacts and integrate Azure DevOps in the GitHub flow
- Discover real-world CI/CD scenarios with Azure DevOps

DevOps Culture and Practice with OpenShift

Tim Beattie, Mike Hepburn, Noel O'Connor, Donal Spring

ISBN: 9781800202368

- Implement successful DevOps practices and in turn OpenShift within your organization

- Deal with segregation of duties in a continuous delivery world

- Understand automation and its significance through an application-centric view

- Manage continuous deployment strategies, such as A/B, rolling, canary, and blue-green

- Leverage OpenShift's Jenkins capability to execute continuous integration pipelines

- Manage and separate configuration from static runtime software

- Master communication and collaboration enabling delivery of superior software products at scale through continuous discovery and continuous delivery

Packt is searching for authors like you

If you're interested in becoming an author for Packt, please visit authors. packtpub.com and apply today. We have worked with thousands of developers and tech professionals, just like you, to help them share their insight with the global tech community. You can make a general application, apply for a specific hot topic that we are recruiting an author for, or submit your own idea.

Share Your Thoughts

Now you've finished *Feature Management with LaunchDarkly*, we'd love to hear your thoughts! Scan the QR code below to go straight to the Amazon review page for this book and share your feedback or leave a review on the site that you purchased it from.

https://packt.link/r/1-800-56297-7

Your review is important to us and the tech community and will help us make sure we're delivering excellent quality content.

Index

www.ingramcontent.com/pod-product-compliance
Lightning Source LLC
Chambersburg PA
CBHW062107050326
40690CB00016B/3241